P9-ECJ-472

Femtocells: Design & Application

About the Authors

Joseph Boccuzzi, Ph.D. has over 20 years of professional experience with companies such as Eaton Corp., Motorola, AT&T Bell Labs, Cadence Design Systems, Morphics, and Infineon Tech. Dr. Boccuzzi is currently a Principal Scientist with Broadcom Corp. designing 3G (WCDMA, HSDPA, HSUPA) and 4G (LTE) wireless cellular communication systems. He is an architect leading the system design for future-generation cellular phones, including baseband, RF, and multimedia. He has represented Broadcom at the 3GPP Home NodeB (i.e., femtocells) standards meetings. Joseph has authored *Signal Processing for Wireless Communications,* published by McGraw-Hill/Professional (2007). He is a recognized expert in signal processing and consumer product implementations, has taught wireless communications courses at Polytechnic Institute of NYU, and offered training courses all over the world. He has written numerous technical papers, holds 20 domestic and international patents, and has made public presentations to various organizations worldwide.

Michael Ruggiero also has 20 years of professional experience, working for companies such as AT&T Bell Labs, Lucent Technologies, Tellium, and Sonus Networks. During his career, he has worked on product and service development for ISDN, ATM, IP Routing, MPLS, VoIP, and IMS technologies. Mike is currently a Distinguished Member of the Technical Staff at Sonus Networks developing call processing and IMS protocol software for carrier-grade products deployed by mobile as well as fixed-line service providers around the world. He holds an MSEE degree from the New Jersey Institute of Technology and has represented AT&T in the development of industry-accepted protocol conformance suites. Mike is an expert in the implementation of telecommunication protocols and products used by today's major service providers.

Femtocells: Design & Application

Joseph Boccuzzi
Michael Ruggiero

New York Chicago San Francisco
Lisbon London Madrid Mexico City
Milan New Delhi San Juan
Seoul Singapore Sydney Toronto

The McGraw·Hill Companies

Library of Congress Cataloging-in-Publication Data

Boccuzzi, Joseph.
Femtocells : design & application / Joseph Boccuzzi, Michael Ruggiero.
p. cm.
ISBN 978-0-07-163358-1 (hardback)
1. Femtocells. 2. Wireless LANs—Standards. I. Ruggiero, Michael, 1964-
II. Title.
TK5105.78.B63 2010
621.382'1—dc22 2010036808

McGraw-Hill books are available at special quantity discounts to use as premiums and sales promotions, or for use in corporate training programs. To contact a representative, please e-mail us at bulksales@mcgraw-hill.com.

Femtocells: Design & Application

1234567890 QFR QFR 109876543210

ISBN 978-0-07-163358-1
MHID 0-07-163358-8

Sponsoring Editor
Wendy Rinaldi

Editorial Supervisor
Patty Mon

Project Manager
Vastavikta Sharma, Glyph International

Acquisitions Coordinator
Joya Anthony

Copy Editor
Robert Campbell

Proofreader
Claire Splan

Indexer
Claire Splan

Production Supervisor
George Anderson

Composition
Glyph International

Illustration
Glyph International

Art Director, Cover
Jeff Weeks

Cover Designer
Jeff Weeks

To my wonderful loving family, Ninamarie, Giovanni, and Giacomo, for being supportive during this project.

—Dr. Boccuzzi

To my lovely wife, Phyllis, and precious daughter, Kristen Marie, for their encouragement and support.

—Michael Ruggiero

Contents at a Glance

Contents

Introduction

As a cellular standard evolves, there are many variables that must be optimized. A few of these variables are directly related to higher throughput, increased user capacity, and in general improved system performance. The optimization techniques have involved higher-order modulation, improved multiple access (in both time and frequency), more powerful error correction codes, cell size reduction, and more. We believe the trend of reducing the inter-cell site distance is the correct solution leading to the femtocell. Femtocells are low-power cellular access points that operate in licensed spectrum to connect standard mobile devices to a mobile operator's network using either DSL, cable, or fiber broadband connections.

The topics covered in this book will provide the reader with a sound foundation in the technical issues designers will be faced with when designing femtocells. This book is intended to be used for graduate-level courses in engineering as well as to serve as a reference for engineers and scientists designing femtocells.

Chapter 1 provides an introduction to the femtocell concept. We discuss competing options of delivering wireless VoIP services in the home. Specifically, we compare the IEEE 802.11 WLAN footprint to that of a cellular femtocell. A discussion of the impact of the handset design will cover cost, size, and power consumption. You will also find a short discussion of the market evolution and the trends that are leading to the use of femtocells.

Chapter 2 presents an overview of the cellular candidate radio access technologies to aid deployment of the femtocell. Here we introduce the 3G Wideband CDMA standard. This will cover the physical layer and some procedural activities. A discussion follows on how packet data services will be handled by the introduction of High Speed Packet Access (HSPA). We then give an evolutionary path for HSPA by discussing high-order modulation and MIMO. Finally, we summarize the 3GPP Long-Term Evolution (LTE) standard, emphasizing certain highlights.

Chapter 3 offers a brief but up-to-date review of the present state of the standards. It introduces the indoor propagation channel and compares path loss models commonly used in designing link budgets, while also providing example results. We review possible home deployment concerns and issues, such as femtocell placement, handling multiple users, and handoffs as well as offer a more detailed discussion of the status of the 3GPP standards effort in defining the system and performance requirements. And continuing along these lines, we provide an overview of the proposed gateway interface to the cellular service provider's networks when femtocells are deployed. Here we present discussions on the standards performance requirements and introduce the 3G Wideband CDMA standard. Finally, we summarize the 3GPP Long-Term Evolution (LTE) standard, while emphasizing certain highlights.

Chapter 4 provides the foundation of understanding for a system and network designer by describing the various approaches to femtocell network architecture. Architecture and scenario diagrams show possible deployment options. Registrations, call establishment, call release, and handoff scenarios are included. We also investigate the signaling and media protocols implemented by the different femtocell network architectures.

Chapter 5 provides an overview of IP telephony. Various VoIP signaling protocols are compared in order to educate the reader and provide the baseline information on why certain industry choices were made. Also described are the advantages and disadvantages of the various protocols. Call flow scenarios illustrate how the various VoIP signaling protocols that exist today compare.

Chapter 6 expands on the numerous options available to the users with respect to the video and voice codec standards, their capabilities and quality/bandwidth differences. The implementation fundamentals of media encoding are illustrated. Also described is the mechanism to carry voice and video over an IP network that is used in today's mobile and wire line networks.

Chapter 7 addresses the security concerns that will inevitably need to be overcome as the femtocell deployment becomes global. Included are comparisons between security levels, network complexities, and resource needs. Implementation specifics on the most common solutions to security threats are provided.

Chapter 8 presents various design considerations, such as delay, jitter, and Quality of Service (QoS). The chapter describes different QoS issues for IP as well as RAN-related protocols and techniques and also offers solutions for managing QoS in an IP-based network in a multimedia context.

Chapter 9 presents the 3GPP IP Multimedia Subsystem (IMS) network architecture. It discusses the various components of an IMS network, including their architecture and the practical responsibilities these network components satisfy. It further describes how the various network components integrate to offer a complete multimedia solution, illustrates detailed call flow scenarios, and describes how VoIP traffic is managed to provide a reliable service. A migration path for delivering VoIP from wire line to wireless is included.

CHAPTER 1

Femtocell Design

What is a femtocell? In the broadest sense, we can use the following definition: a femtocell is: a *low-power* base station communicating in a *licensed spectrum,* offering improved *indoor coverage* with increased *performance;* functioning with the operator's *approval;* offering improved voice and broadband *services* in a *low-cost, technology-agnostic* form factor. Here we have purposely stressed specific key descriptions to convey our message. With the intention of operating indoors, the femtocell will transmit with low power in an authorized frequency band. One of the many benefits of operating in an authorized frequency band is that the operator has the sole rights to utilize it. Hence, the operator controls who communicates in that band and can guarantee a certain level of QoS to all who are involved in occupying the private band.

Providing indoor coverage can be a difficult task, especially due to the propagation path loss of the outer walls of the premises as well as the inter-floor loss. These losses can aggregate to a considerable amount, thus making high-speed 3G data access indoors extremely challenging. Relying on a base station physically located a few kilometers away in distance is not necessarily the best method to effectively deliver high-speed data services to an indoor user—especially since these high-speed data services typically have lower progressing gain and/or use higher-order modulation, such as 64-QAM, to arrive at the high-throughput performance.

The small coverage footprint coupled with the friendly indoor propagation environment will create an atmosphere of high SNR to provide improved performance to support multimedia services at a reasonable price target. Finally, the specific RAT used to provide this feature is operator dependent.

More specifically, the femtocell concept entails using a low-power base station; a cellular phone; and broadband Internet access such as XDSL, cable, or fiber-to-the-home (FTTH). In the residential case all traffic would be routed through the home's ISP connection. This concept is used not only to extend and provide cellular service but also to encourage other applications. The femtocell is sometimes called a personal base station (PBS) or Home NodeB (as referred to in the 3GPP standards body) [1].

This chapter will provide an introduction to the femtocell concept. We will discuss the impact on the complexity of the handset design, specifically with respect to cost, size, and power consumption. We will include a summary of the market evolution and the trends leading to the use of femtocells. Typical usage scenarios will be exercised such as home, friend's home, or party scene. We will also describe some expected applications.

Incorporating the femtocell into a home environment or small-office scenario will open a wide variety of opportunities. Traditionally, the home wireless applications have been less complex, with the exception of WLAN-related options. However, home cordless phones, wireless remotes, etc., have been not only less complex, but also easy to use and feasible for cost, risk, and other reasons. Placing the femtocell into the home will allow users to benefit from the many wide-ranging and highly complex multimedia applications available within the cellular handset sector. As time progresses, it is easy to point out the increased complexity and computing power within these cellular phones, which confer on them bragging rights as being one of the most complex consumer devices in the home. In fact, cellular phone manufacturers are moving their business plans to provide wireless applications to their respective handset platforms, such as Apple, Nokia, Google, and Microsoft.

1.1 The Femtocell Concept

Year upon year cellular service providers struggle to plan for subscriber growth. In order to be prepared for this inevitability, service providers analyze various cell site deployment options. In heavily congested areas the solution has followed a theme to reduce the inter-site distance and provide micro- and even pico-cellular service.

While providing superior system quality of service (QoS) performance, improving cellular coverage is absolutely pertinent, although it can be a daunting task when one tries to satisfy not only the outdoor and highly mobile user but also the indoor and leisurely mobile user. The wireless user will encounter a vastly different experience due to the physical nature of the propagation phenomenon.

It is well known that the lower frequency bands have better propagation characteristics than the higher frequencies and will allow signals to penetrate buildings to reach the indoor users. Moreover, the lower frequency bands improve the link budget, thus allowing the use of higher-order modulation, lower processing gains, etc., which results in higher data throughput to the user. This is part of the reason for the almost absolute about-face from the technology providers racing toward the higher frequency bands to their attempting to revive the lower-frequency bands such as 450 MHz and 700 MHz.

The femtocell or personal base station concept is realized when a cellular service provider places a base station in the home to not only provide better indoor coverage but also to alleviate traffic from the public macrocells. Hence as a user enters his or her home, the cellular phone will recognize the presence of the femtocell and then register to it. This will alert the public macrocell that any further communication to this user will be via the home ISP network. In this case, your cellular phone can behave as a traditional cordless phone; in other words, in addition to its typical cellular traffic, it will now see the traffic from the home usage.

The Femto user is still accessible by the cellular service provider but has freed up resources in the public macrocell that can now be used by additional users that are physically located outdoors. In doing so, the service provider must allow access into their private core network to provide the capability of sending user traffic to the home. This access is provided in the form of a gateway, specifically a femtocell gateway. This provides a dual benefit. First the network operator can now alleviate a fraction of their backhaul traffic to the ISP network. This freed-up capacity will be easily consumed by new users entering the network. The second benefit is to the end user—a higher data rate link can now be established to your phone. Now here is where it gets exciting: a higher data rate will ignite an influx of creative applications to be written for target cell phones.

In Figure 1-1 we show a sample network overview of the femtocell deployment. The homes are expected to have a broadband modem

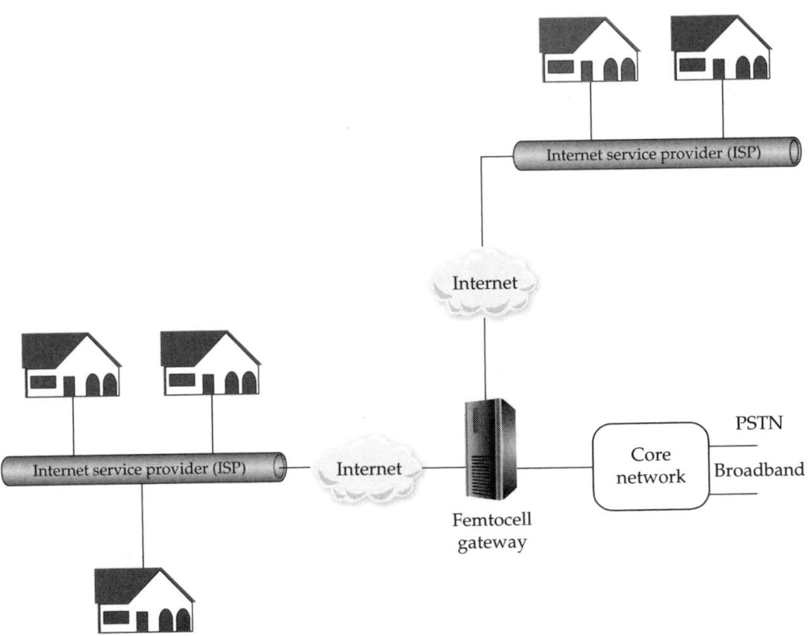

Figure 1-1 Architecture overview of a femtocell network

connection (i.e., XDSL, cable, or fiber) through their Internet service provider (ISP) to the Internet. The cellular specific data will be funneled through the femtocell and enter the femtocell (mobile) gateway for access back into the cellular network. For the Third-Generation Partnership Program (3GPP) network, the gateway would interface to the core network; this interface is called Iu-h.

We have also shown the cellular core network accessing the public service telephone network (PSTN) for voice services and a broadband interconnect for data services. As the core network evolves into a packet network, all traffic will be IP based, thus allowing for a convergence of services.

Cellular phones are sold for operation within specific frequency bands, since these phones are meant to operate in private frequency bands for particular cellular service operators. What this means is that the femtocell will not be allowed to transmit in regions where the service provider doesn't have service nor the rights to that particular frequency band. The owners of these licensed frequency bands are responsible for ensuring emissions satisfy the respective regulatory requirements. Hence, knowing the geographic location of this femtocell is extremely important. This is one particular reason that the femtocells have GPS capability: in order to report back to the cellular service provider the exact location the user is intending to power *on* the femtocell. This will supply the service provider with control needed to restrict the femtocell's operation. Moreover, knowledge of the geographic location is also used to support emergency services, as well as lawful interception and a host of other reasons. We wish to quickly follow up by noting that GPS is not the only method available to provide location information; service provider IP addresses and other means are also available. We believe a combination of all of these will lead to an accurate and satisfying experience.

We have thus far not described the use of the public ISM frequency band, meaning WLAN is not included in this definition. What we have discussed thus far is a system employing the cellular (or wide area network, WAN) RAT, and not one from the personal area network (PAN) such as Bluetooth or the local area network (WLAN or WiFi) such as IEEE 802.11, although VoIP traffic over WLAN service is increasing, especially with the introduction of the iPhone. We believe these WAN and PAN services will continue to coexist, since they solve specific issues and provide services that are sometimes orthogonal in nature to one another.

Cellular service providers have paid exorbitant prices for the regional licensed spectrum; hence, they have the legal rights to use the spectrum. Moreover, from one service provider to the next the spectrum properties (bands, regulations, etc.) differ not only nationally, but also from one country to another. For naming purposes, the network used in the femtocell will be called the *private* network, while the network used for typical cellular communications will be called the *public* network. This naming convention will be used to aid the descriptions that follow.

In Figure 1-2, we show the possible combinations of the private and public networks. Here the public macrocell is shown by a single, large oval coverage area. Within this area we have specifically drawn four Home NodeBs (HNB), using the 3GPP nomenclature. They are identified as follows:

- HNB-A is geographically located near the macro-NodeB.
- HNB-B is located near the cell fringe.
- HNB-C is located in an area where cell coverage is spotty.
- HNB-D is co-located with HNB-B.

Please notice we haven't differentiated between the private and public network user equipment (UEs), since they should be able to seamlessly travel within their respective networks. Next we will discuss each of these geographic locations.

The HNB-A position is located near the public, high-power NodeB. If the macrocell is using the same frequency band as the private cell, then the downlink of the private and public networks can see an increase in interference. As a result of this increase in the downlink interference, UEs located within the HNB-A coverage will see a degraded downlink from the public macrocell. When moving indoors, however, the public macro-signal becomes attenuated by the factors already discussed, whereas the indoor private femtocell signal is increased. Here the outermost wall is used in a positive manner and extremely welcome. This wall will not only attenuate the signal entering the home from the public cell but also attenuate the signal exiting the home from the private cell to help reduce downlink interference within the private and public networks, respectively.

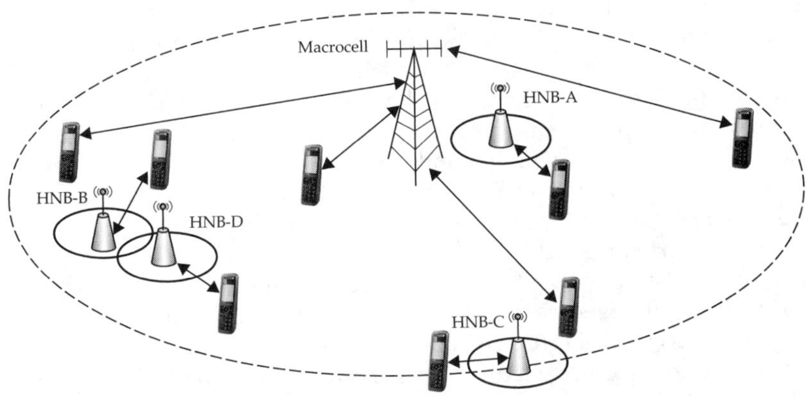

Figure 1-2 Femtocell (Home NodeB) interactions with macrocell

The HNB-B position is located near the cell edge. If the macrocell is using the same frequency band as the private cell, then we would generally expect to have smaller downlink interference due to the increased propagation loss on the downlink. In this scenario the use of the femtocell has increased the downlink throughput due to the better SNR of the femtocell compared to the public macrocell offering. However, the UE is located at the macrocell edge and will transmit with higher power than the UE associated with the HNB-B. Here the HNB-B uplink will experience a larger rise in interference, since the two UEs are not both associated with the HNB. Here interference mitigation techniques should be applied carefully so as not to allow an increase in the HNB transmit power to overcome this shortcoming, since closed loop power controlled systems have the potential to be unstable (or closely approach it, thus requiring QoS intervention).

The HNB-C position is located at the cell fringe, where we have included the possibility that cell coverage can be nonexistent. If the macrocell is using the same frequency band as the private cell, then the downlink interference is expected to be small, but the uplink can be significant, depending on the location of the public UEs. In this case the femtocell has increased the cell coverage and also improved the available data rates to the end user.

The HNB-D position is located near HNB-B, where we have purposely needed to include interference generated by neighboring femtocells operating in either the same frequency or adjacent frequency. Here both femtocells experience uplink and downlink interference from the macrocell. We must note for multiple cases, however, that the interference from HNB can now deteriorate performance of users in the macrocell. Hence users operating near a few hundred HNBs, for example, may experience some sort of performance degradation within close proximity. The 3GPP standard's group is working diligently to minimize this occurrence.

Although this single-cell example was used to convey the potential interference the femtocell would need to overcome, similar issues arise when multicell deployment scenarios are considered. Finally, when the adjacent frequency bands are considered, interference issues still exist and should be carefully planned. Let's consider the apartment complex scenario where many users are operating within the building and potentially the adjacent complex. Users associated with the macrocell can easily have degraded performance not only outside but also indoors due to the rise in co-channel interference (CCI).

To fully support the femtocell concept, a few components need to be defined: personal base station, handset, ISP, gateway, and cellular network. Figure 1-3 provides an example of a single femtocell architecture. Here we have a single UE communicating to the HNB, which has a coverage area that can extend slightly beyond

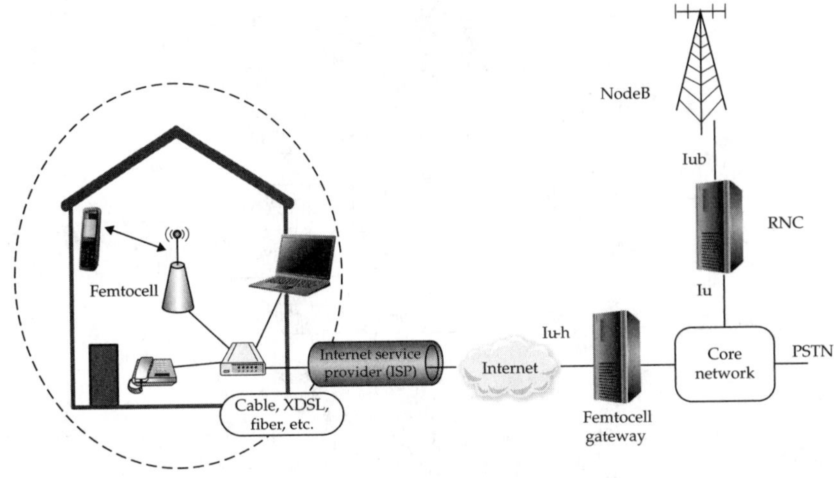

Figure 1-3 Femtocell architectural components

the home premises. This HNB plugs into a broadband modem to access the Internet. Access back into the cellular network is available through the femtocell gateway. We have also drawn upon the fact that many users currently get their dial tones via the IP packet network. Also, home desktop computers are connected via a broadband modem to high-speed Internet access.

1.1.1 Market Overview and Direction

At the time of writing of this book, there are many companies providing a partial or complete lineup of femtocells solutions. Here is a small sampling: chipset providers such as Picochip; service providers such as AT&T, Verizon, Orange, Telecom Italia, Telefonica, and T-Mobile; and manufacturers such as Samsung, Motorola, and Nokia.

A high-level block diagram of the implementation components of the Home NodeB is shown in Figure 1-4. A HNB can be a separate "box" that would essentially connect to the broadband modem, or the modem functionality can be integrated.

Homes will also need telephone lines for already available corded and cordless telephones; hence, the RJ11 connection is available. Home security systems can use this method as well. The broadband modem can connect to a WiFi access point or router to allow other devices such as a desktop computer to access the Internet. Also, many homes have WiFi service; hence, this can also be a separate box or have this functionality integrated as well. Many options exist, and we believe the variable worth optimizing is cost when considering initial deployments.

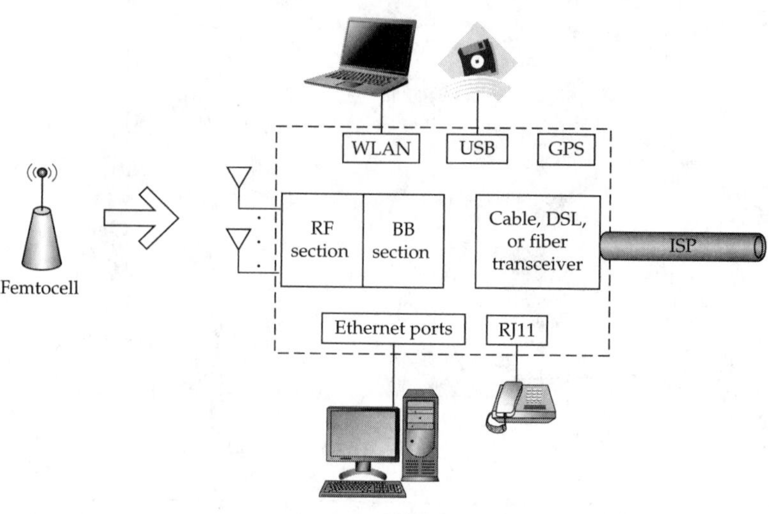

Figure 1-4 Home NodeB components

A long-term strategy should also be considered. One obvious solution would be the integration of WLAN protocols such as 802.11a, b, g, and n. We expect and hope the integration would provide an overall cost advantage. One point worth mentioning is the rate of evolution of the various technologies integrated. For example, the WLAN standards have been evolving at a faster pace than cellular. If not carefully studied unnecessary limitations can be imposed on consumer product roadmaps.

1.1.2 Insights into the Cellular Roadmap

The intention of this section is to give the reader insight into the evolution of the cellular radio access technologies (RATs) such as GSM, WCDMA, and LTE. These standards are evolving in order to reduce end-to-end latency, provide packet services capability, increase the data throughput, increase user capacity, etc.

In Figure 1-5 we plot the data rate (in Bps) versus the spectral efficiency (in Bps/Hz) for the various RATs. There is a clear trend toward increasing the spectral efficiency as the standards evolve. We have chosen to display the reference deployment information when comparing the overall cellular roadmap.

Spectral efficiency can be used to further calculate user data rate, throughput, or even system capacity, which then justifies our reasoning for choosing the performance metric. This figure shows us WCDMA offered some performance improvement over GSM and further improvement when HSPA was deployed. Here Higher-Order Modulation (HOM), as well as other techniques, was used to produce this improvement. Even further improvements can be seen with the introduction of Long-Term Evolution (LTE).

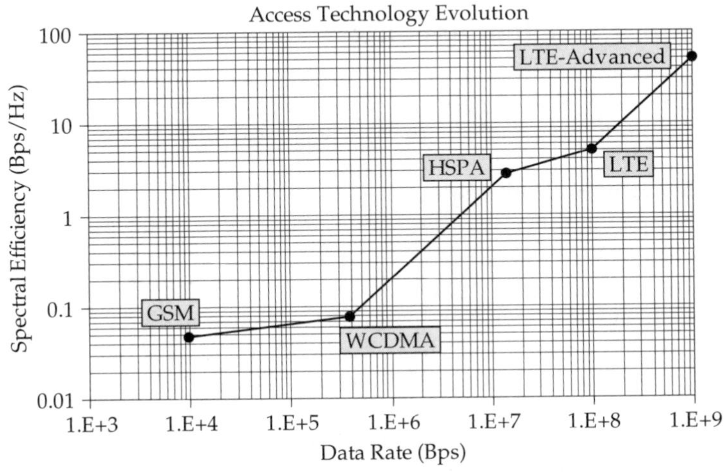

FIGURE 1-5 Cellular radio access technology evolution

In Figure 1-6 we compare the respective evolutionary paths of GSM, WCDMA, and LTE. For example, GSM is evolving to GPRS, EDGE, and EDGE Evolution. This migration is shown along the leftmost straight line in the figure. Next WCDMA is evolving to HSDPA and HSDPA-Plus using higher-order modulation schemes, MIMO, and dual-carrier techniques. Finally, LTE has evolved to LTE-Plus using increased MIMO techniques and then to LTE-Advanced.

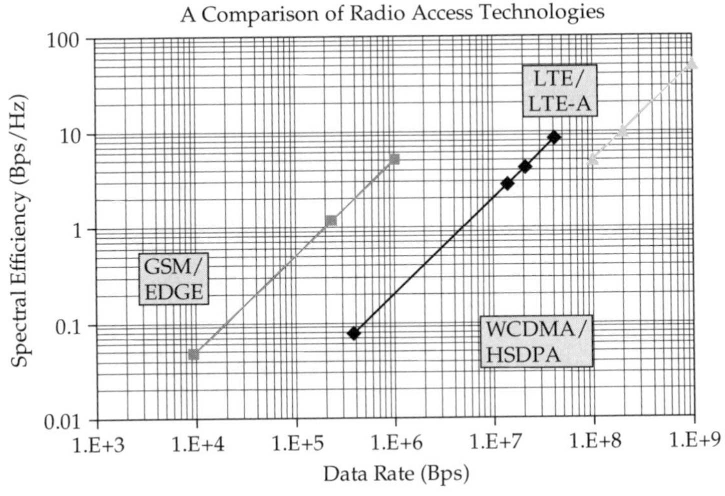

FIGURE 1-6 Comparison of cellular radio access technologies

1.2 Femtocell System Benefits

The femtocell concept will bring forth many questions on performance, use cases as well as benefits. Next, a list of expected benefits to not only the user, but also the network provider will be reviewed [2].

1.2.1 Typical Deployment

In this subsection we will outline the items to consider for a typical deployment.

The femtocells are allowed to transmit in certain frequency bands. Service providers have purchased large chunks of frequency bands for large sums of money. These frequency bands are referred to as *licensed* bands. This operational scenario is very different from the well-known WLAN case, which happens to use the unlicensed bands (ISM bands of 2.4 GHz and 5 GHz, etc.).

It is commonly accepted that a significant portion of cellular phone calls are started from within buildings. Hence we should consider cases when phone calls (or traffic sessions) initiate indoors and then eventually may need to be carried over to the outdoor public macro-network. In this case a hard handoff is made from one cell to another. Similar reasoning can be applied in the opposite direction, where phone calls were originated outdoors and then enter the femtocell coverage area. It is expected that the phone call (or traffic session) would hand off into the private network in order to gain the expected performance improvements.

Increasing the number of macrocells in a network is expensive. Pushing part of the network build-out to the end customer can alleviate some of the operator costs; however, for a deployment to be considered successful, millions of customers should be using the service. Hence the operators should have a good plan to support many HNBs and their associated UEs.

How Will the User Configure the Femtocell?

After turning *on* the femtocell, do we rely solely on the GPS signal to obtain the femtocell geographic coordinates to be transmitted back to the cellular network to identify if this femtocell is indeed in a valid area? This may very well be the case, but it is worth mentioning that the access IP address and/or cellular neighbor cell list can also be used to help with HNB authentication and registration. We believe a combination of the above or others will lead to an accurate indication of location.

How Will We Address Open vs. Closed Networks?

Next, we briefly describe the concept of open and closed networks. Open networks essentially allow anyone near the femtocell access to it, while closed networks do not allow any UE to connect to the femtocell unless the UE has permission from either the home user or network provider. This enables the home owner and the service provider to control what users can access their network. Initial deployments are expected to be closed networks. They allow for a more controlled version of the initial deployment. Moreover, one should also consider the situation of when the indoor users have been addressed, the next logical step would be to allow the HNB coverage zone to extend to the outer sections of the home premises to accommodate pedestrian traffic. This will surely require open network deployments. Moreover campus and/ or small business style scenarios would benefit from an open network.

How Do We Add More Users to the Femtocell in the Home?

Will users have a special user interface to identify that a home network is nearby and to ask for permission? This will be a truly personal experience.

When you come home and hand off to the femtocell, then your cellular calls and home calls should be sent through your home femtocell. Is there an audible signal or user interface (UI) icon that indicates the call is from the cell or POTS?

The typical femtocells have emerged with the capability to support up to four simultaneous users within the home or small office. This will support both voice and data traffic. It is envisioned that as time progresses the capacity of the femtocell will increase with the continued introduction of advanced receivers for both the handset and the NodeB. Last but not least, having a better understanding of the interference problems and in turn the interference mitigation techniques will also help toward increasing capacity.

Control should ultimately go to the service provider, since it is their spectrum being used. Having a femtocell that will automatically configure itself (scrambling codes, frequency, etc.) is desirable and an integral component to the interference mitigation. The optimal approach, from the service provider's perspective, is to not perform additional radio planning and network dimensioning every time a new femtocell is sold. In fact, a network that is self organizing (SON) is highly desirable.

Migrating from legacy RATs, such as GSM, to newer systems using LTE is interesting. Various options can be considered here. First, the application of LTE to femtocells can be an excellent introduction to the RAT. However, this means that the UE must be available and not only support LTE, but also WCDMA and the expected GSM services as well. As with any new technology introduction, initially the power consumption will be more than is desirable, and size may be an issue. Once system measurements are collected, we would be able to observe improvements in the network that should result in better overall customer satisfaction.

1.2.2 Advantages to the Femtocell User

In this section we outline the envisioned femtocell benefits to their users. Placing the femtocell indoors in either a home or small office environment will unleash a great potential to the user. The following is a list of expected benefits to the femtocell users.

- **Simple deployment** It is advantageous to simplify the user's involvement in setting up the femtocell deployment in the home or small office environment. This reduction in complexity should assuredly lead to an increase in the probability of a successful deployment.
- **Increased user throughput** With the user physically close to the femtocell, a reduction in block error rate (BLER) is certainly possible, as well as expected. This would increase the average data throughput to the user, opening up possibilities. As is well known, users closer to the NodeB will enjoy higher throughputs than those closer to the edge of the cell.
- **Improved indoor coverage** Placing the femtocell indoors would alleviate the concern to include an additional 10–15 dB of loss into the system link budget. Hence, depending on the location of the home or small office with respect to the public macrocell, in-building coverage can be an issue. Moving a low-power NodeB indoors will extend the cellular footprint of the service provider, and the user would benefit from this. From the public macro-perspective, propagating through walls is undesirable, but required. However, from the point of view of the femtocell, using the walls to keep the interference inside and attenuate it outside is an extremely good property.
- **New applications** Having a convergence of mobile and home-based devices will lead to a wide variety of new applications for the user. These applications are applicable to both home and small office environments. As phones continue to allow for open applications development, unlimited personal and professional uses arise. For example, to provide the ability to use the femtocells to support electronic medical devices (such as electronic bandaids and EKG meters) and allow such information to be routed to the medical community should accelerate this nascent field.
- **Reduced power consumption** The user will no longer need to transmit with high power, since the femtocell is located near the user. This will translate into smaller current drain from the battery, resulting in longer standby times and increased talk times.
- **Enhanced multimedia/IP services** Allow the user to have an enhanced experience with videos, home services, phones, computers, etc.
- **Improved voice quality** Having the user so close in proximity to the femtocell will provide a better communication link. This will allow the use of better-sounding, higher–data rate speech vocoders to be used.

- **Security** Since the femtocell's connection is via the public ISP and traffic will be routed to the private cellular network, the user must authenticate himself to the network; the service provider can use IPsec. The users can rest at ease knowing that personal or professional information will be secure.
- **Improved customer satisfaction** With benefits of improved indoor coverage and high throughputs, service providers expect the user to have a satisfying experience.
- **Business** With the current direction of cellular phone applications, it is conceivable that owners of these private femtocells can have specific applications that would allow such users special privileges and could possibly bypass the public macrocell for certain scenarios. As a related comment, the near-term femtocells accommodate up to four simultaneous users.

Among other areas of concern to femtocell users, let us outline the following:

- **Billing** This has the potential to be devastating to the femtocell user. Here the user will receive a separate broadband connection bill, possibly from the cable or telephone company rather than the cellular service bill. Moreover, the cellular service bill will now have additional costs of using the femto service. Service providers will be faced with difficult decisions on whether to charge users on a call origination basis or use another more creative approach such as a flat monthly rate. We believe the basic operating premises should be low cost and clarity. Piling on additional costs, charges, and fees to femtocell users has the strong potential to slow adoption of this service. Similarly, delivering a confusing payment plan can be disastrous.
- **Measurements** For the scenario when many femtocells are deployed, the UE connected to the macrocell will not only be able to make measurements on the macrocells, but also the many femtocells it encounters. There is a potential negative side effect that the UEs in this particular femtocell area will report back many measurements to the public network that may or may not be useful for certain configurations. The UE can potentially make many more neighboring cell measurements, which can in turn overwhelm the core network with the reporting of such measurements.
- **Home resources** Depending on the level of integration, a few boxes may be needed to complete one home. In doing so, more of your home's real estate area has been taken up, not to mention the additional electric power bills to supply energy to those boxes.

1.2.3 Advantages to the Network Provider

In this section we outline the envisioned femtocell benefits to the service providers. Placing the femtocell indoors in either a home or small office environment will open up great opportunities to the service providers. The following is a list of expected benefits to these operators.

- **Increased revenue** An increase in the average revenue per user (ARPU) would occur when both the number of users in the network increases and the monthly revenue increases. Having the femtocell capability with the service provider would hopefully attract additional users. Moreover, the femtocell application can be viewed as an additional feature/ service, not currently covered under your contract, in which case the service providers would be able to charge the user for use of the service.
- **Reduced cost** As the service providers strive for increased coverage and capacity, the network complexity grows, and unfortunately with this so does cost. Shrinking the public cell size from macro to micro and then to pico has its capacity benefits, but it adds to network costs (deployment, maintenance, backhaul, recurring expenses, site rental, etc.). Using an already-available network such as the Internet, cost reduction is certainly feasible, given the communications move to the all-IP-based structure. Having said this, it is expected that this cost savings will eventually make its way to the end user.
- **Increased capacity** Having users currently connected to the public macro-network move to the private femtocell would open up physical resources (frequency, time slots, scrambling codes, etc.) so that others can connect with the public cell, while still maintaining the present customer base. This will increase the numbers of users in the system overall.
- **Improved indoor coverage** Service providers have for the longest time struggled with coverage, especially indoor coverage using a public macrocell NodeB. Placing the femtocell indoors will extend the coverage region of the provider, since the additional 10–15 dB required to penetrate the building walls and floors would no longer be needed. In fact, under ideal conditions, the service provider should take these no-longer-needed dBs and use them to provide a higher cell throughput.
- **Enhanced services** Using the user's access to the Internet, the service provider can reveal enhanced services tailored around the user's phone, cellular network, and home, hopefully to improve the user's efficiency and quality of life.
- **Compete with other convergence technologies** Currently, the most effective and commonly used RAT to access the Internet from the home environment is via the WLAN.

As cellular technology data rates improve, this would hopefully allow users to also use the cellular network to transfer data, in addition to the voice and multimedia traffic expected.

- **Product differentiation** When considering the stiff competition among cellular service providers, the option or capability for this femtocell service along with the cell phones is a great product differentiation.
- **Improved customer satisfaction** With benefits of improved indoor coverage and high throughputs, service providers expect the user's expectations to be satisfied.

It is interesting to consider how ISPs will react to cellular service being carried over their networks if they don't reap the financial benefits. A few service providers that provide cellular and broadband service to homes will find that this easily supports the quadruple-play service model. For the rest of the world, a different picture emerges. Broadband providers have collaborated to bring forth WiMax services, while almost simultaneously outdoor WiFi hot spots have appeared offering low price points to attract and potentially keep wireline customers.

Among other areas of concern to the cellular service providers, let us outline the following:

- **Business** With the current direction of cellular phone applications, it is conceivable that owners of these private femtocells can have specific applications that would allow such users special privileges and could possibly bypass the public macrocell for certain scenarios. As a related comment, the near-term femtocells accommodate up to four simultaneous users (with long-term targets in the neighborhood of 20 users). The concept of a public femtocell will be interesting.
- **Measurements** For the scenario in which many femtocells are deployed, the UEs will report measurements not only from the macrocells, but also from the many private femtocells. This may present a need to increase the measurement processing capability of the network.

1.3 Handset Impact

In this subsection we will address various hurdles, issues, and questions related to the impact of the femtocell scenario to the complexity of the already-complicated handset. The handset has been riding the waves of improved battery life, reduced size, and enhanced features (such as multiple cameras and displays), and it would be a shame if this femtocell capability would be disruptive to this progression rather than an enabler. Whenever possible, the associated issues surrounding the femtocell will be discussed in order to provide a more complete picture of the system issues. Our intention is to also provide a non-polarized viewpoint to the reader.

The first question we would like to address is: Does the phone need to be modified?

We believe the best match would be to have a cellular phone that is femto-capable. By this we mean it is aware of the special femto environment and can therefore perform as well as support specific features that are amenable to the home or small office application. Having a small icon letting the user know it is now in its home network is a powerful awareness tool. After all, this has been done for 3G and is expected to be used for LTE and so on.

Some femtocell adoption strategies have been to load a web-based application on a home computer that would allow additional users to join your home private network as guests. This would mean for every new visitor, you would need to walk over to the computer to allocate the necessary permissions. It is conceivably a better situation if the home user's cellular phone would have this application embedded into the handset as well, to give more flexible control over not only adding new users, but also removing them as necessary. Moreover, it would allow the home user to exercise some preference over which users are allowed and which are not. Recall the early femtocell models allow up to four users, which is extremely limited when the home is hosting a family gathering or having a celebration.

When we view this femtocell scenario, the following additional features should be located on the phone:

- **Display icon** Having an indicator that tells the user whether she is on the private cell or the public cell is important. If any features associated with the private network have a financial impact, the end user would like to verify its operation.
- **Audible beeps** If cellular phones will be used in the home as some evolved cordless phone, a habit must be overcome of looking for the second phone. When a cellular phone is in the private cell and receives phone calls from the private network for the home, a special ring tone, audible beep, or other signal should be used to alert the user of the type of phone call she is about to answer.
- **User interface for femto-control** Having the private UEs with control capability is an excellent feature. Because of the above-mentioned interference concerns, it would behoove the home or small office private network to allow certain users to connect to its network. Similarly, there may be times when certain UEs should be forced off the private network when others of higher priority return to the network.
- **Preference setting for hand-offs** Just as cellular phone users can now select 2G versus 3G preferences, it would be beneficial to offer the same capability to the femto users as well.

- **Femtocell quality indicator** For the scenarios when many femtocells are deployed, the femto user may want to have an interference or quality picture (snapshot in time) of the present situation. This parallels the advanced uses of WLAN scenarios we currently see in the market.
- **Applications** The femtocell is an excellent opportunity to mix certain home applications that would otherwise potentially use WLAN technology. Applications that utilize multimedia in the home are extremely valuable and meant to increase the user's experience.

Other features will doubtless find their place.

1.3.1 Complexity Discussion

The impact of the deployed frequency band to handset complexity is specifically visible when receive as well as transmit diversity is used. The physical location of the antennas becomes more important as we lower the frequency band of operation. This arises from the spatial separation requirement to obtain close to uncorrelated waves at each of the receiver antennas. For certain multipath environments, this minimum separation is on the order of half of a carrier wavelength.

The initial deployment of the LTE RAT is expected to be around the 700 MHz frequency band range. However, worldwide service demands do vary.

The frequency synchronization or frequency stability is standardized by 3GPP and chosen to be 0.25 ppm [3]. The HNB manufacturer has a few options in order to achieve this stringent goal. One is to use the GPS signal itself to derive a stable clock. Since the GPS is envisioned to be used in the femtocell configuration and authentication phase, this seems like a reasonable approach. Please be aware that any satellite-based location service works best when line of site is available. In other words, placing the GPS receiver close to a window or open space would be favorable. A second technique is for the femtocell to have a UE receive option that it can use to demodulate the macrocell DL in order to lock onto the 0.05ppm clock stability of the macrocell. In this case the manufacturer would then schedule periodic measurement intervals where the HNB would monitor the public cell for frequency stability reasons [4]. In fact, this can be carried further to also demodulate the broadcast channel to extract useful network information from the surrounding public cells such as scrambling codes, etc. This information can be used in an interference mitigation algorithm to control and/or reduce neighboring femtocell interference.

Various other options exist such as combinations of Internet clock synchronization and purchasing of tighter crystals such as Temperature Compensated Crystal Oscillator (TCXO) as well as other more expensive components. But please be aware that crystals have an aging specification that should not be ignored if the user is expected to have the femtocell for more than a few years. In any case, all techniques have a certain cost associated with them and can be used in the product differentiation aspect of the femtocell. More details surrounding the 3GPP performance requirements will be supplied in later chapters.

1.3.2 Dual-Mode Designs (WLAN + Cellular)

With the tremendous success of the WLAN technology, it is only natural to suggest a network or a service that would make use of this not only technically but also from a financial perspective. Many homes, coffee shops, airports, parks, and other public places already offer such free and/or pay WLAN access. In order to address this comment, we will briefly look at the potential of a handset with not only the cellular (i.e., femtocell) capability, but also the WLAN as well.

Figure 1-7 shows a simple block diagram of the so-called dual-mode handset operating in the expected usage scenarios.

Our discussion will center on the following items:

- **Complexity** The additional handset complexity will be seen from both the hardware and software perspectives. First, the cellular platform will need to support the WLAN (i.e., IEEE 802.11a/b/g/n) features. The exact protocol a dual handset must follow to scan for available WLAN or public cellular networks is not fully defined by either standards group. In fact, this would deserve special attention, since this would affect performance. Moreover, the additional measurements required for supporting handoffs from one RAT to another, etc., should be considered. This complexity can lead to longer design, development, and manufacturing time lines, all leading to longer time to market.
- **Cost** Additional chipsets are required to support this WLAN functionality besides the cellular chipsets. These chipsets should make use of the combination of other RATs such as

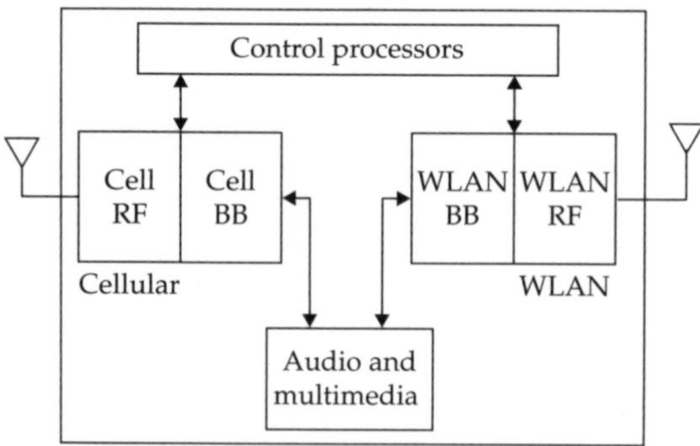

FIGURE 1-7 Dual handset usage scenario

GPS, BT, FM, and DVB-H. The addition of such features can preclude entry into the low-tier market sector. To some extent this is already being seen with the adoption of the iPhone on the Asian markets.

- **Battery life** Based on the upper-layer protocols chosen, having the handset periodically scan for either cellular (both public and private) or WLAN service will create a drain on the battery current. This has the potential to significantly reduce talk time and stand-by time in the dual-mode cellular handset, especially when both of these operations are occurring frequently. This would negatively impact the marketing of such cellular phones and hurt overall competition and product differentiation. There are many potential solutions to this conundrum. For example, if GPS is used the coordinate can be collected to trigger the WLAN searching capability. Similarly the indication that a femtocell is near can also be used to assume a WiFi access point is not far away.
- **Availability** When compared to the entire handset population, the dual-mode capability has a limited selection of handsets. As time progresses, this limitation will surely be lifted and users will have more options. From the handset perspective, as the adoption of Open OS continues to proliferate, the dual-mode landscape will surely change.
- **Mobility** Having the dual-mode handset will allow users to operate in environments that support either WLAN or cellular. We use the phrase "or" because an original motivator in dual-mode handsets was reducing the cost to the end user; however, with the continually dropping costs of service providers, this motivation may need to be revisited, albeit on a case-by-case basis. Moreover, with the improved coverage of cellular service in airports, libraries, etc., the landscape is continually changing.
- **Performance** Special attention should be paid to the overall use cases, especially to concurrent scenarios. Having a phone simultaneously perform the following functions is challenging: receive an MM message via the cellular system, use GPS to update/track position on a map, have WLAN download a web page, and have BT send audio to the earpiece. The processor speeds on a cell phone are lower than laptops and desktop computers (albeit the gap is becoming more and more narrow every few years). Hence the users should not expect the same throughput performance as observed on a computer.

A service that currently exists and can be viewed by some as a complimentary service to femtocells involves WLAN technology. Service provider A can have WCDMA femtocells, while Service

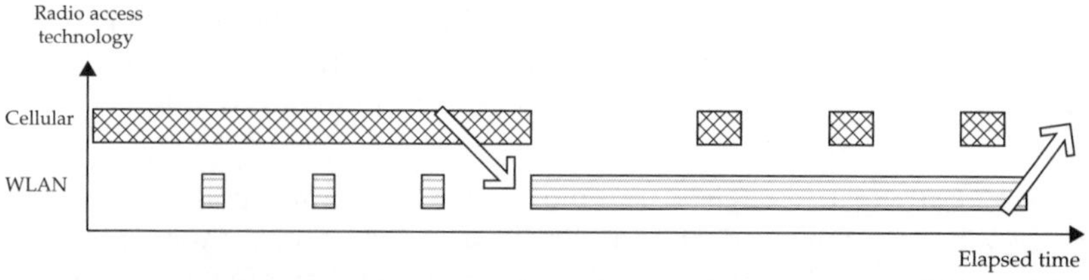

FIGURE 1-8 Dual handset monitoring example

provider B also has WLAN. Service provider B would be able to support handoffs to and from the cellular network.

Figure 1-8 shows an example where the UE is operating on the cellular service and has the WiFi periodically awakened to scan for WiFi. Once WiFi has been found and/or other criteria have been met, then UE can connect to the WiFi service and have cellular periodically awakened to scan for public and private cells. As discussed previously, the periodicity as well as the exact triggering mechanism is considered to be part of the product differentiation picture. It suffices for now to note that the scanning frequency has a significant impact on the overall battery life.

1.4 Femtocell Applications

In this section we will review some of the common expected usage cases for the femtocell users.

1.4.1 Home Usage Models (Femtocell Architecture Overview)

We next provide an introduction to the Iu-h interface between the HNB and the HNB-GW in the 3GPP standard. Figure 1-9 provides a femtocell network architecture diagram along with the functionality partitioning suggested. Where HNB is used to denote Home NodeB, HNB-GW reflects the Home NodeB gateway operator, CN is the core network, and H-UE is the UE connected to the Home NodeB.

Near each entity we have listed several functions performed in each block. Pushing the radio resource management (RRM) functions to the edge of the network is a strategy to not only reduce system cost but also support increasing user capacity. Here we also see the gateway services to authenticate and register not only the UE, but also the HNB.

1.4.2 Femtocell Protocol Overview

The interface defined between the Home NodeB and Home NodeB gateway is named Iu-h [5], [6], and [7]. An overview of the protocol is provided in Figure 1-10, where the user and control planes are highlighted. In this configuration both circuit-switched (CS) and packet-switched (PS) data streams are supported.

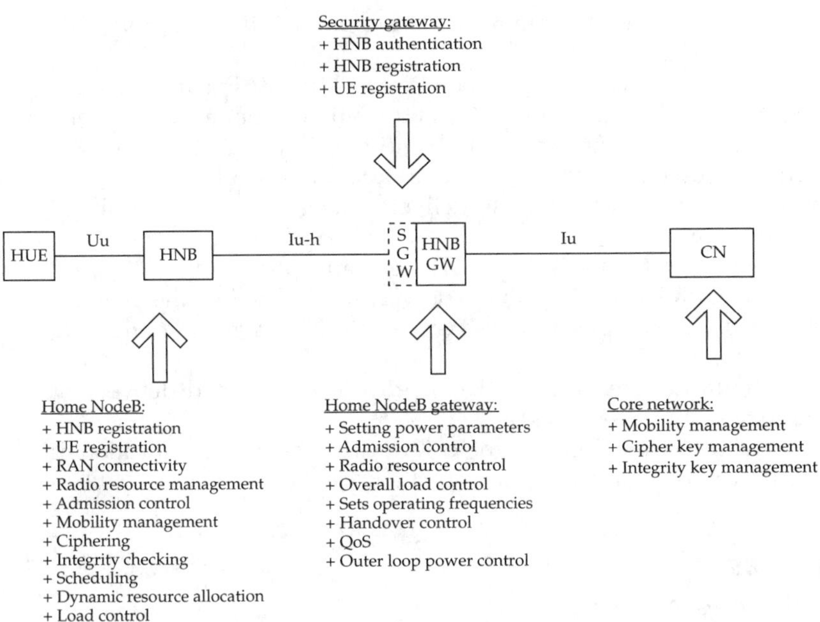

FIGURE 1-9 Femtocell network architecture

The protocol elements shown in Figure 1-10 will now be defined.

The UTRAN functions consists of radio access bearer (RAB) management, radio resource management (RRM), Iu link management, mobility management, and security, as well as other functions.

The HNB functions consist of HNB registration management, UE registration to the HNB, Iu-h management, etc.

The Radio Access Network Application Protocol (RANAP) is used in the control plane of the stack between the UTRAN (RNC) and the CN, specifically the Iu interface. This can be viewed as the control plane signaling protocol.

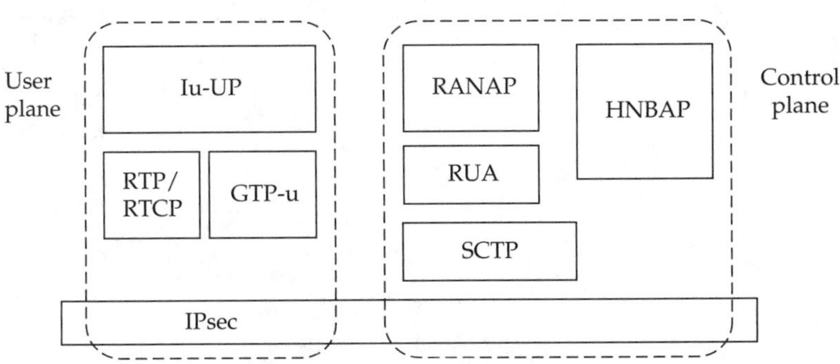

FIGURE 1-10 Home NodeB Iu-h protocol overview

RANAP user adaptation (RUA) supports HNB and HNB-GW, error handling, etc.

The Home NodeB Application Protocol (HNBAP) supports HNB registration, UE registration, HNB and HNB-GW communication, etc.

The Stream Control Transmission Protocol (SCTP) is a transport protocol (used by SIP) to provide secure and reliable transport for next-generation networks. It delivers datagrams where multiple streams are allowed.

The Iu-User plane is the Iu-UP user traffic plane.

The GPRS Tunneling Protocol (GTP) is defined as the protocol between GPRS support nodes (GSNs), for both signaling and data transfers.

The Real-Time Protocol (RTP) provides end-to-end delivery of data for real-time services.

The Real Time Control Protocol (RTCP) provides an indication of the transmission and reception of data carried by RTP.

References

[1] www.3gpp.org

[2] S. R. Saunders, et al., *Femtocells: Opportunities and Challenges for Business and Technology*, J. Wiley & Sons, 2009.

[3] 3GPP Technical Specification 25.104: "Base Station (BS) radio transmission and reception (FDD)," 2010.

[4] US Patent #6,370,157, J. Boccuzzi and W. Lieu: "Automatic frequency control for a cellular base station."

[5] 3GPP Technical Specification 25.468: "UTRAN Iuh Interface RANAP User Adaption (RUA) signalling," 2010.

[6] 3GPP Technical Specification 25.469: "UTRAN Iuh interface Home Node B (HNB) Application Part (HNBAP) signalling," 2010.

[7] 3GPP Technical Specification 25.467: "UTRAN architecture for 3G Home Node B (HNB); Stage 2," 2010.

CHAPTER 2

3G and LTE Radio Access Technologies

In this chapter we will present an overview of the candidate cellular radio access technologies (RAT) to aid deployment of femtocells or Home NodeBs as coined by the 3GPP standards body. This introduction will cover the physical-layer details of WCDMA, HSPA, HSPA Evolution, and LTE. Here general operating procedures, time slot structures, required transport channels, and available data rates (i.e., throughputs) will be discussed. It is also important to review the underlying assumptions of each RAT as they pertain to the delivery of packet data services, since the next-generation networks (NGNs) will be all IP based. As the successful deployment is defined as one that has global acceptance, we must provide information regarding the proposed deployment spectrums to be used.

We will present a system-level perspective through the use of network architecture block diagrams that focus on packet- and circuit-switched services. We will show the evolution of the network architectures and the respective positions of the protocols within. This will reveal how certain network improvements will be achieved through the reallocation of certain functions.

In discussing the packet evolutionary path of HSPA, we need to identify the benefits of high-order modulation (HOM) and impact of using multiple antennas (MIMO), as these two techniques will be used extensively in successful LTE deployments worldwide. This chapter will provide the foundation for the femtocell system analysis presented in Chapter 3.

2.1 WCDMA Overview

WCDMA continues to enjoy its global success as a result of the careful planning and precise execution of the 3GPP standards group, equipment manufacturers, and service and content providers. Improvements to this RAT standard are made in releases. Figure 2-1 provides an overview of the additional functionality awarded to this 3GPP system as a result of each release.

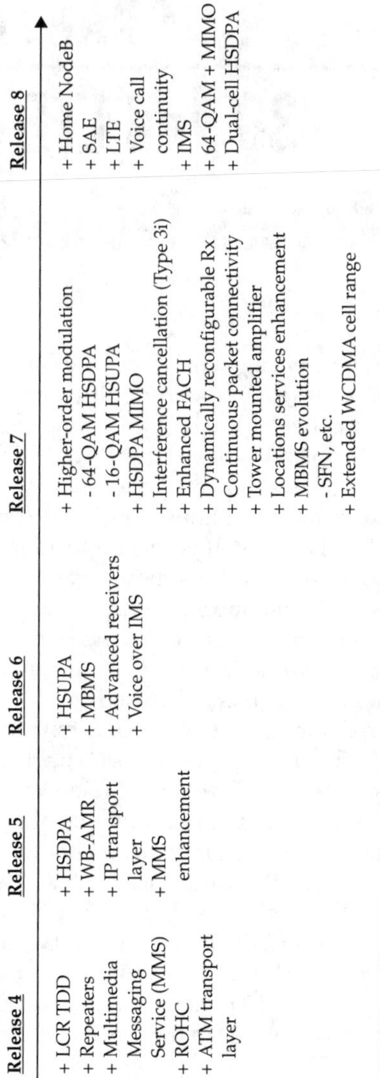

Release 99	Release 4	Release 5	Release 6	Release 7	Release 8
+ WCDMA + R99 radio bearers + ATM transport layer	+ LCR TDD + Repeaters + Multimedia Messaging Service (MMS) + ROHC + ATM transport layer	+ HSDPA + WB-AMR + IP transport layer + MMS enhancement	+ HSUPA + MBMS + Advanced receivers + Voice over IMS	+ Higher-order modulation - 64-QAM HSDPA - 16-QAM HSUPA + HSDPA MIMO + Interference cancellation (Type 3i) + Enhanced FACH + Dynamically reconfigurable Rx + Continuous packet connectivity + Tower mounted amplifier + Locations services enhancement + MBMS evolution - SFN, etc. + Extended WCDMA cell range	+ Home NodeB + SAE + LTE + Voice call continuity + IMS + 64-QAM + MIMO + Dual-cell HSDPA

FIGURE 2-1 The 3GPP standard release functionality roadmap

The WCDMA aspect of the 3G cellular standard was introduced in 1999 and is called Release 99. After this initial release, the standards group decided to change the release naming convention, and thus later versions were renamed starting from Release 4.

The 3GPP releases show after the introduction of WCDMA, upgrades were included to address packet access capability by moving dedicated and common-channel traffic to shared channels, as exemplified by HSPA. This effort is required to support IP data traffic as well as Voice over IP (VoIP) telephony services.

In an almost simultaneous effort, the focus of the 3GPP standards group was placed on performance improvement not only at the cell edge, but across the entire cell. This improvement will be visible with the introduction of advanced receivers applicable to not only the UE, but also the NodeB. These new receivers can, depending on the propagation channel conditions, deliver several dBs of improvement to the system. A point worth mentioning at this moment is that overall performance gain will only be had once a significant amount of users within the cell have these advanced receivers; otherwise, it becomes more of a personal enhancement.

2.1.1 Frequency Bands

The 3GPP-based RAT specifications are targeted to support both frequency division duplex (FDD) and time division duplex (TDD) operations. Paired spectrum is used for FDD scenarios with the UE transmitting only in one band and while listening to another band of frequencies. When transmitting and receiving occur at the same time, this is called full-duplex communication. Similarly, when transmitting and receiving do not occur at the same time, this is called half-duplex communication.

The unpaired spectrum can be used for broadcast scenarios as well as for TDD. In the latter case, the UE transmits and receives in the same frequency band. Since the frequency band is shared between the uplink and downlink, transmission and reception does not occur at the same time. This option has the potential to be very attractive in regions where paired frequency band access is not possible.

The 3GPP standard has defined the following 14 frequency bands for global deployment. Table 2-1 shows the uplink and downlink frequency bands necessary for communication.

2.1.2 Network Architecture

In this subsection we will provide a baseline understanding of the WCDMA network architecture, which will establish our reference point as we discuss the evolution of the 3GPP RATs. The WCDMA network architecture is illustrated in Figure 2-2. The network consists of the following components: NodeB, Radio Network Controller (RNC), Core Network (CN), and UE [1].

Operating Band	Uplink Frequencies UE Transmit, NodeB Receive	Downlink Frequencies UE Receive, NodeB Transmit	Regions
I	1920–1980 MHz	2110–2170 MHz	Europe, Asia
II	1850–1910 MHz	1930–1990 MHz	Americas, Asia
III	1710–1785 MHz	1805–1880 MHz	Europe, Asia, Americas
IV	1710–1755 MHz	2110–2155 MHz	Americas
V	824–849 MHz	869–894 MHz	Americas
VI	830–840 MHz	875–885 MHz	Japan
VII	2500–2570 MHz	2620–2690 MHz	Europe
VIII	880–915 MHz	925–960 MHz	Europe, Asia
IX	1749.9–1784.9 MHz	1844.9–1879.9 MHz	Japan
X	1710–1770 MHz	2110–2170 MHz	Americas
XI	1427.9–1452.9 MHz	1475.9–1500.9 MHz	Europe, Americas
XII	698–716 MHz	728–746 MHz	Europe, Americas
XIII	777–787 MHz	746–756 MHz	Europe, Americas
XIV	788–798 MHz	758–768 MHz	Europe, Americas

Table 2-1 3GPP Frequency Bands for FDD Operations

Here NodeB is a term used to represent a base station transceiver (BTS) and is responsible for physical-layer signal processing. These responsibilities consist of modulation, demodulation, FEC encoding and decoding, spreading and despreading, Control + Dedicated + Shared Channel aggregation, etc. In Figure 2-2 we have drawn the NodeB to consist of three sectors; here the antenna patterns are used to collectively cover 360 degrees surrounding the NodeB site. A salient point worth mentioning here is that we have assumed a single frequency in this discussion so far. However, sites using multiple sectors, such as six, or multiple frequency bands, such as two, are not uncommon, and hence we feel we should at least mention them.

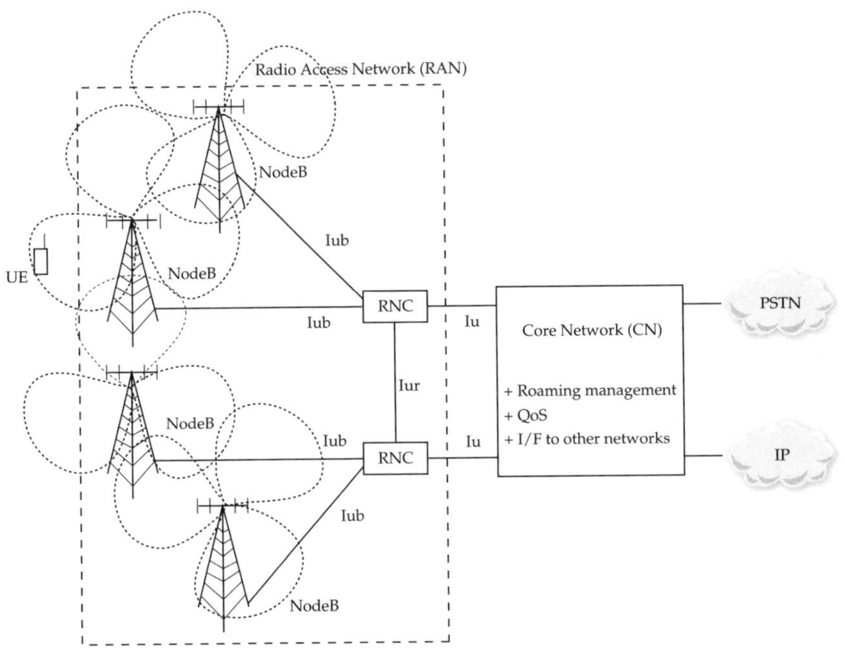

FIGURE 2-2 WCDMA network architecture

A varying number of NodeBs can be connected to a RNC, depending on cell site deployment, available BW, estimated capacity, and desired QoS. The RNC is responsible for the radio resource management of the sites, congestion and admission control, handover control, ciphering and deciphering, open loop power control, administering QoS solutions, etc. Essentially, the RNC is performing the allocation of resources. In order to connect the voice or data sessions to their desired destination, the RNCs will be interfaced to the Core Network (CN). The CN has the overall Radio Access Bearer (RAB) management.

The overall management of the voice or data sessions can be described through the use of a protocol architecture block diagram as shown in Figure 2-3. The physical layer (PHY) is located in the NodeB, while the MAC, RLC, and RRC are located in the RNC.

The Radio Resource Control (RRC) controls the radio resources. The Packet Data Convergence Protocol (PDCP) is used for header compression of IP packets and is defined only in the PS domain.

The Radio Link Control (RLC) protocol is used for segmentation and re-assembly of IP packets, ARQ protocol, concatenation,

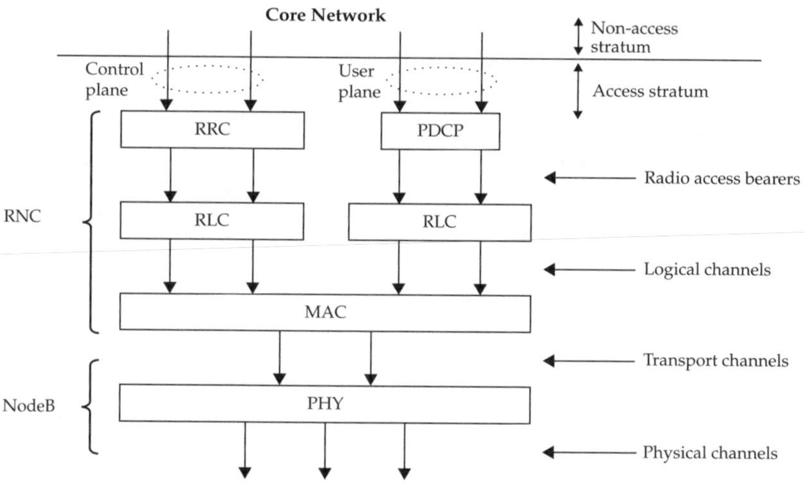

Figure 2-3 WCDMA protocol architecture

in-sequence delivery, and retransmission of both users and control data. The RLC operation consists of three modes:

- **Transparent Mode (TM)** There is no overhead inserted into the RLC layer. This is applicable to the CS voice services.
- **Unacknowledged Mode (UM)** No RLC re-transmissions are expected; used for delay-sensitive services such as VOIP.
- **Acknowledged Mode (AM)** This has RLC retransmissions for services that require all the data packets to be delivered successfully to the higher layers.

The Medium Access Control (MAC) maps the logical channels from the RLC into transport channels for the PHY layer. The timing instances of which blocks of data are sent to the PHY from the MAC are called Transmission Time Intervals (TTIs). Since resources are shared between not only users but also applications, this TTI defines FEC parameters such as FEC interleaver depth and coding techniques.

The PHY, MAC, RLC, and PDCP layers are configured by the Radio Resource Control (RRC) protocol. The RRC is used for handover control, call admission, etc. This is accomplished by providing the RRC with many measurements taken from various layers in order to provide the RRC with the best possible current viewpoint, in order to make the best possible decisions.

As shown in Figure 2-3, the PHY layer resides in the NodeB, while the RLC, RRC and MAC all reside in the RNC (this functionality partitioning is valid for the WCDMA scenario). It is easy to witness potential system latencies embedded in this architecture, since data must be exchanged between physically different components as you traverse upward, and downward, on the protocol architecture.

The protocol architecture just described is separated into Access Stratum (AS) and Non-Access Stratum (NAS). The AS carries the user and control signaling that corresponds to the access technology. Examples of AS signaling are power control, channel allocation, and handover control. The NAS carries the user and control signaling that essentially is independent of the access technology. Examples of NAS signaling are call setup, call control, authentication, and registration. The original reason the network and protocol architecture were partitioned into these two parts was to have the capability to independently evolve each one from the other.

The Radio Access Network Application Part (RANAP) provides the signaling between the RAN and the CN. Some of the RANAP protocol functions are as follows:

- Relocating the SRNC.
- Overall Radio Access Bearer (RAB) management.
- Enabling the RNC to request RAB release.
- Enabling the RNC to request the release of all Iu connection resources.
- Controlling overload in and resetting of the Iu interface.
- Sending the UE Common ID (NAS UE identity) to the RNC.
- Paging the user. This function provides the CN for capability to page the UE.
- Transport of NAS information between UE and CN.
- Controlling the security mode in the UTRAN. This is used to send the security keys (ciphering and integrity protection) to the UTRAN.
- Data volume reporting function. This function is responsible for reporting unsuccessfully transmitted DL data volume over UTRAN for specific RABs.
- Uplink Information Exchange. This allows the RNC to transfer/request information (such as MBMS) to the CN.
- Overall MBMS RANAP control. This function allows for MBMS RAB management (setting up, updating, and releasing the MBMS RAB).

Uplink and Downlink Channels This subsection provides a review of the uplink and downlink channels required for supporting the necessary WCDMA and HSPA functions. More details surrounding the physical channel contents can be found in the 3GPP standards [2–6].

The FDD mode chip rate is fixed at 3.84 MCps. The physical channels are carried in the form of the following physical resources:

Radio Frame	=	10 msec in duration
	=	Consists of 15 time slots
	=	Consists of 5 subframes
	=	Consists of 38400 chips
Time Slot	=	10/15 msec (or 0.667 msec) in duration
	=	Consists of 2560 chips
Subframe	=	2 msec in duration
	=	Consists of 3 time slots
	=	Consists of 7680 chips

Downlink Channel Structures Four types of downlink dedicated physical channels are defined: DPCH, F-DPCH, E-RGCH, and E-HICH.

The downlink frame and time slot structure for the Dedicated Physical Channel (DPCH) is given in Figure 2-4. The DPCH consists of Dedicated Physical Data Channel (DPDCH) and Dedicated Physical Control Channel (DPCCH) time multiplexed into the physical channel. This type of channel carries user data and control information dedicated to a particular user. The transport format combination indicator (TFCI) is used to tell the UE details regarding the transport blocks transmitter. The TPC bit field is a physical-layer control issued to control the UE uplink transmit power control.

The spreading factor (SF) of this channel can vary from 512 down to 4. Since the total number of chips transmitted in a time slot is fixed, the variable SF translates into having a variable number of bits in each time slot field. The time slot formats are well defined in 3GPP Technical Specification (TS) 25.211. When comparing the uplink to

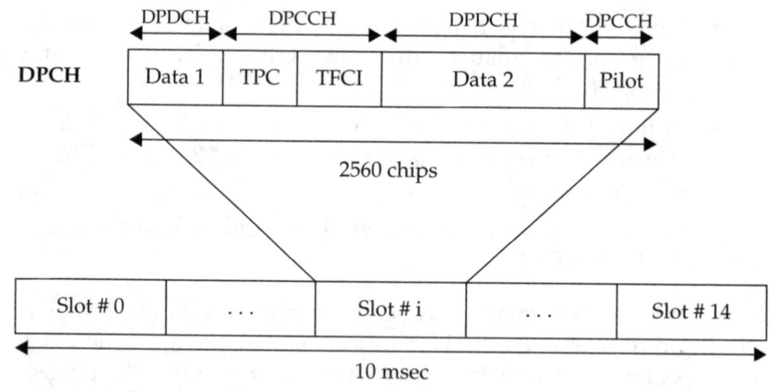

Figure 2-4 Downlink DPDCH and DPCCH time slot and frame structure

the downlink channels, the major difference is on the uplink, where the DPDCH and DPCCH channels are code multiplexed, while they are time multiplexed on the downlink.

The Enhanced-DCH Relative Grant Channel (E-RGCH) time slot and frame structure is given in Figure 2-5. The SF is fixed at 128 and this channel is used to carry the E-DCH relative grants. Here we describe the channel contents, where as the usage model will be more apparent when we present the HSUPA feature of the 3GPP standard.

The relative grant is transmitted using 3, 12, or 15 consecutive time slots, each slot consists of 40 ternary values. The 3 and 12 time slot durations are used on E-RGCH in cells that have E-RGCH and E-DCH in the serving radio link set. The 3 and 12 time slot durations correspond to the E-DCH TTI of 2 msec and 10 msec, respectively. The 15 slot duration is used on E-RGCH in cells that have E-RGCH, not in the E-DCH serving radio link set. The 40 ternary values are generated by multiplying an orthogonal signature sequence of length 40 by the relative grant value, a. The relative grant value has two definitions: $a \in \{-1, 0, +1\}$ in a serving E-DCH radio link set and $a \in \{0, -1\}$ in a radio link not belonging to the serving E-DCH radio link set.

The next downlink dedicated physical channel is the Fractional Dedicated Physical Channel (F-DPCH), the time slot and frame structure for which is shown in Figure 2-6. The SF is fixed at 256 and is used to carry control information for layer 1, specifically the TPC commands. The position of the TPC command within the time slot is variable to accommodate multiple users time multiplexing their TPC commands onto the same physical channel.

F-DPCH is a special case of the downlink DPCCH channel; it contains only the uplink transmit power control commands. This channel removes the need to have a dedicated downlink DPDCH channel, unless it is necessary. A typical scenario is to use the F-DPCH channel with user traffic that is sent over the HS-DSCH channel. One such scenario can be the VoIP application. All downlink data/voice will be sent to the UE via the HS-DSCH, with F-DPCH present only to signal

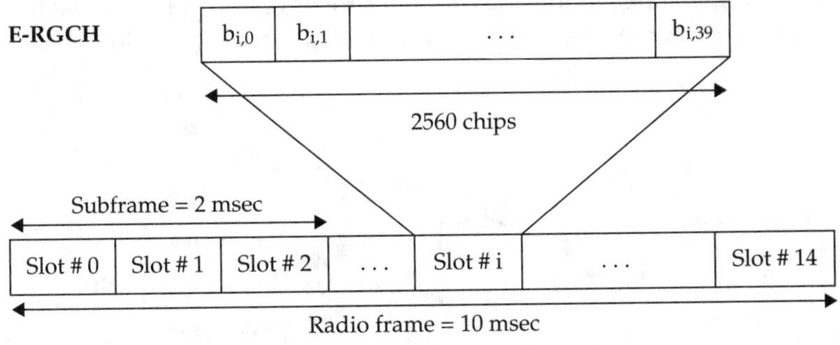

FIGURE 2-5 Downlink E-RGCH time slot and frame structure

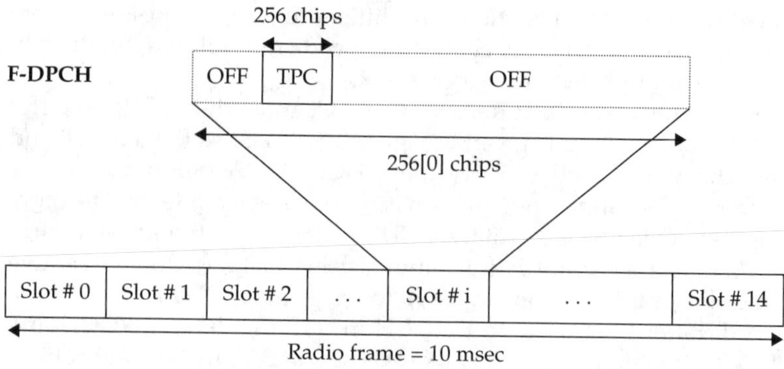

FIGURE 2-6 Downlink F-DPCH time slot and frame structure

power control commands. The uplink channels remain the same, as will be discussed in the next sections, namely DPDCH, DPCCH, and HS-DPCCH. The last point to mention here is that by dramatically reducing the downlink signals, for example, no DPDCH transmissions and a fraction of the DPCCH by using F-DPCH, the downlink interference (both within the cell and outside the cell) is significantly reduced.

Due to the time slot structure, it is conceivable that multiple users can have their F-DPCH channels time multiplexed onto a single physical channel where each user has a multiple of 256 chips time offset. This time-multiplexed scenario can also be used to conserve the usage of channelization codes. We will leave the E-HICH channel discussion until the HSUPA services are discussed in later sections.

The downlink physical channel operations can be summarized as in Figure 2-7, where only a single physical channel is shown for illustrative purposes. The bits to be transmitted are first converted to symbols depending on modulation scheme used. The modulation symbols are then spread by a channelization code, here the OVSF codes are used. This is represented as $C_{CH,SF}$ where the subscript CH stands for channelization and SF stands for spreading factor. Next the complex spread waveform is scrambled by a Gold Code. This is

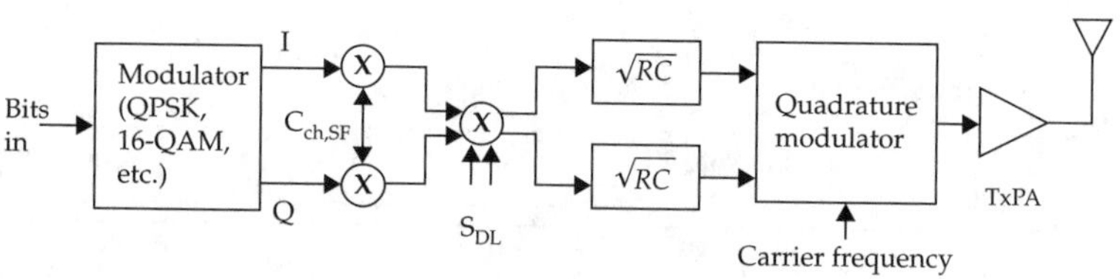

FIGURE 2-7 Downlink spreading and scrambling

represented by S_{DL}, where the double lines indicate a complex multiplication (i.e., real and imaginary). The chips then become filtered and enter the quadrature modulator for frequency translation prior to being amplified by the transmit power amplifier (TxPA).

Next we outline some pertinent information related to these signal processing operations. The modulator maps the even and odd numbered bits to the In Phase (I) and Quadrature Phase (Q) Channels, respectively. The channelization code is time aligned with the symbol time boundary. The scrambling code is time aligned with the P-CCPCH frame. For other channels not frame aligned with P-CCPCH, they will not have the scrambling code time aligned with their physical frames. The pulse shaping filter used is the square root raised cosine (SRC) with a roll-off factor $\alpha = 0.22$. The channelization codes are called OVSF, where they preserve the orthogonality constraint between a user's different physical channels.

Uplink Channel Structures There are five types of uplink dedicated physical channels: DPDCH, DPCCH, E-DPDCH, E-DPCCH, and HS-DPCCH.

The Dedicated Physical Data Channel (DPDCH) physical channel is used to carry the DCH transport channel. The DPCCH is used to carry the layer 1 control information such as TPC command and TFCI and FBI bits. The Uplink frame and time slot structure is given in Figure 2-8 for the data and control channels, DPDCH and DPCCH, respectively.

The SF of DPCCH is constant and equal to 256, while the SF of DPDCH varies from 256 down to 4, depending on the user data rate selected. The duration of the bit fields is clearly provided in the 3GPP Technical Specification 25.211 and will not be repeated here [2]. The pilot bit field contains a known sequence of bits that changes in each

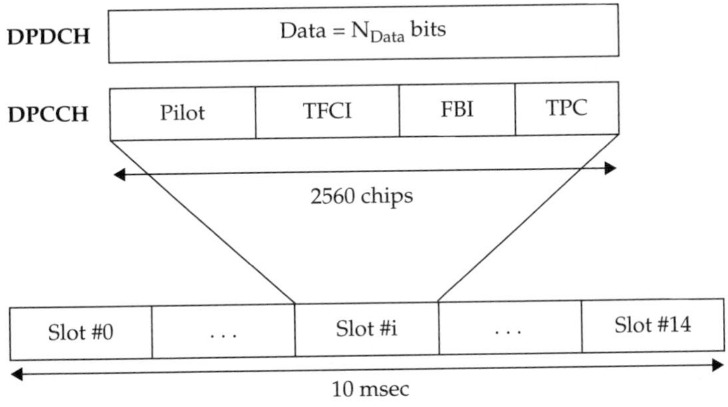

FIGURE 2-8 Uplink DPDCH and DPCCH time slot and frame structure

time slot and repeats every frame. These pilot bits can be used to aid channel estimation for coherent detection, assist in beam forming, aid SIR estimation, perform frames synchronization, etc. The Transport Format Combination Indicator (TFCI) bits inform the receiver about the instantaneous transport format combination of the transport channels mapped to the simultaneously transmitted DPDCH radio frame. The FBI bits are used to support feedback information required from the UE for such application as closed-loop transmit diversity (CLTD). Finally, the TPC bits are used to convey power control commands to adjust the downlink NodeB transmit power. Since these TPC bits are all of the same polarity (i.e., either all 1's or all 0's), is it conceivable that they also can be used to aid the channel estimation or other signal processing algorithms if so required.

The frame and time slot structure for the next two uplink dedicated physical channels is given in Figure 2-9 for E-DPDCH and E-DPCCH. These channels correspond to the Enhanced DPCH support for HSUPA.

The SF of E-DPCCH is fixed at 256, while the SF for E-DPDCH can vary from 256 down to 2. The E-DPDCH is used to carry the E-DCH transport channel, and the E-DPCCH is used to carry the associated control information. The enhanced DPCH channel will be used to send uplink packet data for HSUPA services. Based on certain downlink measurements, the UE scheduler will send appropriate packets on the uplink E-DPCH channels. The number of channels will be discussed in the section "HSUPA Details" later in this chapter.

The last uplink dedicated physical channel to be discussed is HS-DPCCH; its subframe and time slot structure is provided in Figure 2-10.

The SF for HS-DPCCH is fixed at 256 and has a single slot format. The HS-DPCCH carries the uplink feedback signaling related to the downlink transmission of HS-DSCH, supporting HSDPA services.

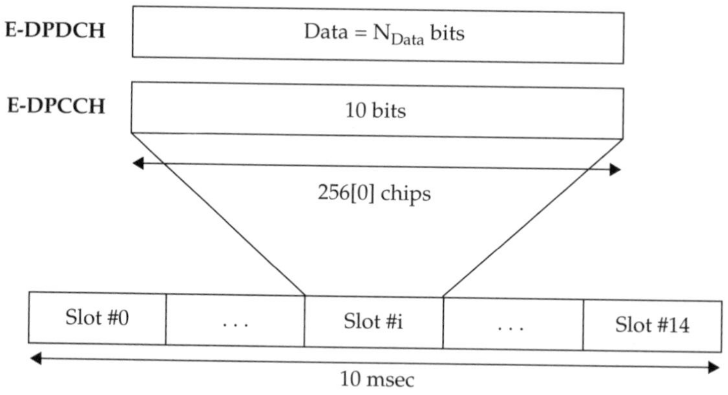

Figure 2-9 Uplink E-DPDCH and E-DPCCH time slot and frame structure

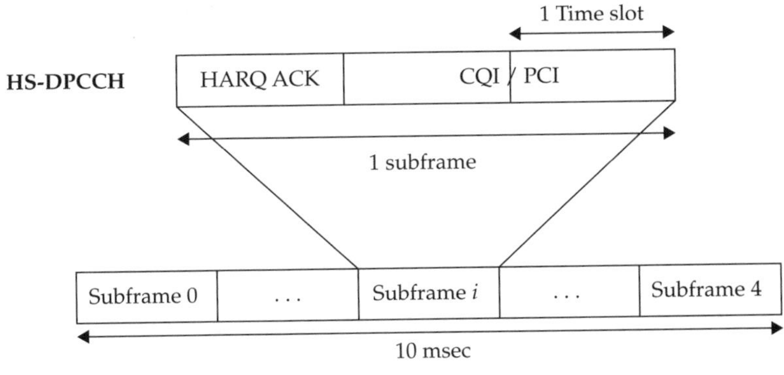

Figure 2-10 Uplink HS-DPCCH time slot and subframe structure

The uplink feedback signaling consists of the HARQ Acknowledgment (ACK/NACK), the Channel Quality Indicator (CQI), and Precoding Control Information (PCI). The first part of the subframe consists of the acknowledgment, while the remaining consists of the CQI information.

The acknowledgment is used by the NodeB as a response to the demodulation of the downlink HS-DSCH packet transmitted. If the HS-DSCH CRC decode is successful, then an ACK is inserted. If the HS-DSCH CRC decode is unsuccessful, then a NACK is inserted. The CQI is a time varying indication of the channel the UE observes. The mapping of this quality to a quantitative indicator is accomplished by sending the NodeB an index into a channel quality table indicating the best possible Modulation scheme, Transport Block Size, and Code Rate the UE can receive at that particular time instant. We will discuss this operation in more detail in the section "HSDPA Details" later in this chapter.

Five uplink physical channels were presented in this subsection. The DPDCH and DPCCH are time slot aligned along with the E-DPDCH and E-DPCCH channels. The HS-DPCCH channel is not necessarily time aligned with these channels due to the uplink and downlink timing relationship required to support HSDPA services. The uplink spreading and scrambling operations can best be described with the help of Figure 2-11 for DPCH channels. In this figure the code multiplexing of DPDCH, DPCCH, HS-DPCCH, E-DPDCH, and E-DPCCH are shown. Each channel has a channelization code sequence and an associated power scaling (i.e., Beta factor). This scaling is used to allocate appropriate uplink resources.

In Figure 2-11 we see that multiple DPDCH channels are possible, up to six to be exact. They are alternatively placed on the I and Q channels. The DPCCH is always placed on the Q channel. The HS-DPCCH is alternatively placed on the I and Q channels, depending

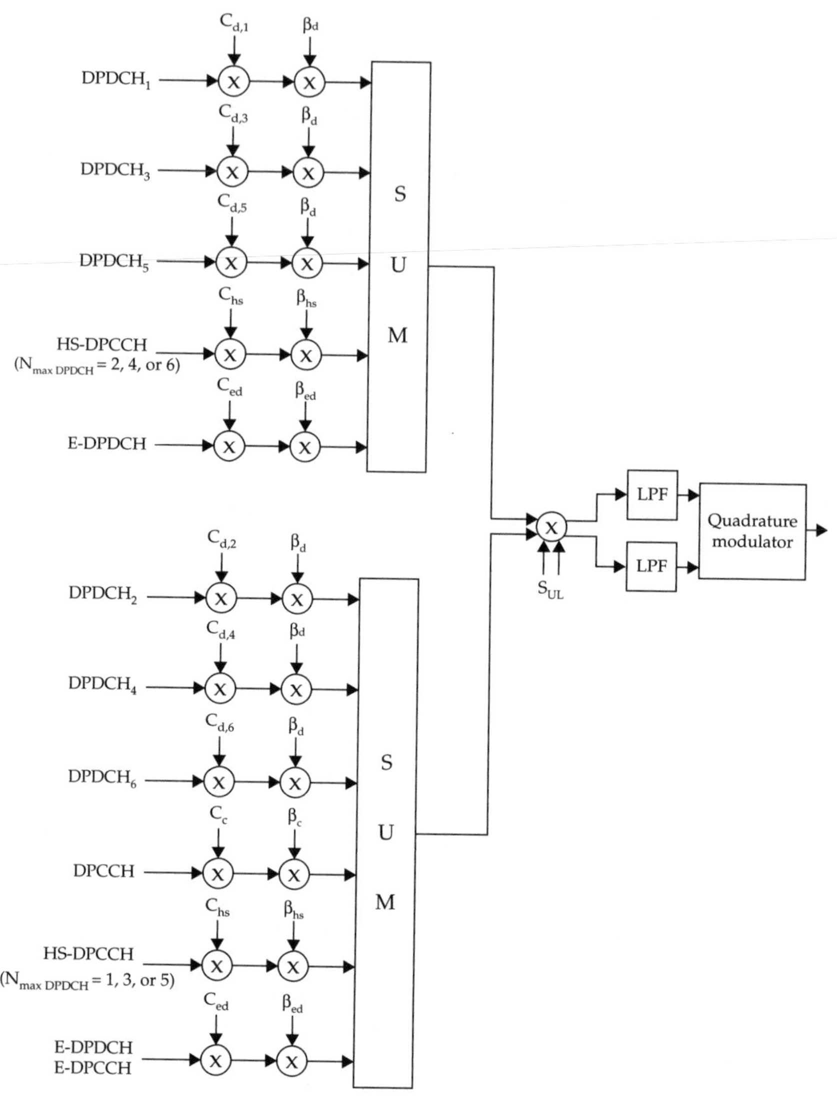

FIGURE 2-11 Uplink scrambling and spreading overview

on the number of multiple channels used. The E-DPDCH is also alternatively placed on the I and Q. Finally, the E-DPCCH is placed on the Q channel. The relative powers between the channels are controlled by the "β" multipliers. The gains, Beta values, are signaled by the higher layers. The DPCCH is always spread by channelization code $C_{ch,256,0}$.

The uplink scrambling codes are assigned by higher layers. The scrambling codes are generated by modulo 2 sum of two binary

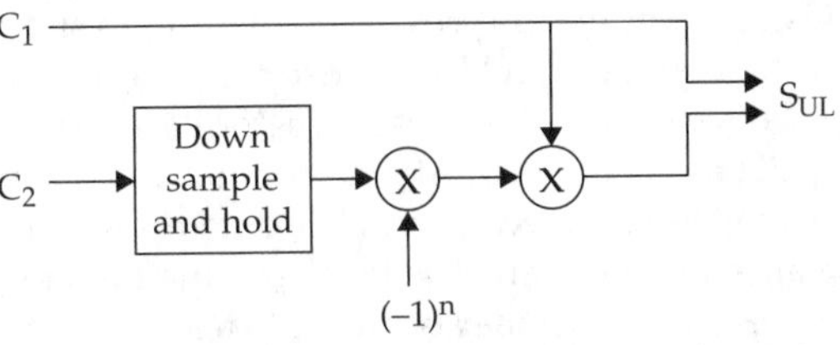

FIGURE 2-12 Uplink scrambling code generation for HPSK modulation

m-sequences generated by two generator polynomials of degree 25. The uplink scrambling code is a complex-valued sequence where the imaginary part, C_2, is a time-shifted version of the real part, C_1. The next part of the scrambling code generation is to create the Hybrid PSK (HPSK) modulation, which is accomplished as shown in Figure 2-12.

The HPSK functionality shown in the figure was introduced to lower the uplink Peak to Average Power Ratio (PAPR). After scrambling, the uplink waveform is called HPSK modulated. Essentially, the number of times the 180-degree phase change is encountered has been reduced, thus having the desirable side-effect of lowering the out-of-band emissions, particularly when a nonlinear device is inserted into the transmission path, such as a power amplifier. In Figure 2-12 the letter n is used to denote the chip number index.

2.1.3 HSPA Overview

In order to accommodate more packet data users, WCDMA was extended to include what is called High Speed Packet Access (HSPA). These enhancements covered both downlink, called High Speed Downlink Packet Access (HSDPA), and uplink, named High Speed Uplink Packet Access (HSUPA) connections.

HSDPA Details

Some of the requirements in order to successfully support packet data services are: increased peak data rates, reduction in latency, increase in capacity, improvement in performance. These requirements are fulfilled utilizing techniques such as HOM, Dynamic Scheduling, Hybrid ARQ, rate control, efficient use of shared channels, resource control. This particular functionality was introduced into the 3GPP standard in the Release 5 version.

A UE will identify its capabilities by reporting to the NodeB its category. The HSDPA UE category identifies the maximum number of parallel HS-DSCH codes it can receive, the minimum inter-TTI interval supported, modulation schemes supported, buffer size, etc. The scheduler is located in the NodeB to make use of this information for the scheduling of users' data traffic.

A transmission channel will be shared in order to support HSDPA. This physical channel is called High Speed Downlink Shared Channel (HS-DSCH). This means that resources within the cell can be dynamically allocated between users as the need is determined. This bursty behavior is well suited for packet data applications. The HS-DSCH physical channel has an SF = 16. Users and data rates are trade-offs as a function of the number of multicodes used. The number of multicodes available for transmission ranges from 1 to 15, which depends on the available resources and the UE capability.

Information needed for the UE to demodulate and decode the HS-DSCH is transmitted over the High Speed–Shared Control Channel (HS-SCCH). Hence typical HSDPA services consist of dual steps: the first is to demodulate the control channel, and the second will demodulate the data shared. In this subsection we will present the physical channels required for HSDPA operations.

High Speed–Downlink Shared Channel (HS-DSCH) The HS-DSCH has a constant SF = 16 and can be scrambled by either the primary or secondary scrambling code. The time slot and subframe structure is given in Figure 2-13. This HS-DSCH channel carries the packet data already described.

The HS-PDSCH shown corresponds to a single channelization code from the set of possible channelization codes reserved for the HS-DSCH transmission. The total number of possible codes is 15. HS-DSCH may use either QPSK or 16-QAM or 64-QAM. Please note the HS-PDSCH does not carry layer 1 control information; all relevant layer 1 information is transmitted on the associated high-speed control channel (HS-SCCH). Moreover, while HSDPA services are provided, the uplink/downlink WCDMA link must be established to maintain a communication link.

High Speed–Shared Control Channel (HS-SCCH) The SF is fixed at 128. The channelization codes used for HS-DSCH are allocated as follows:

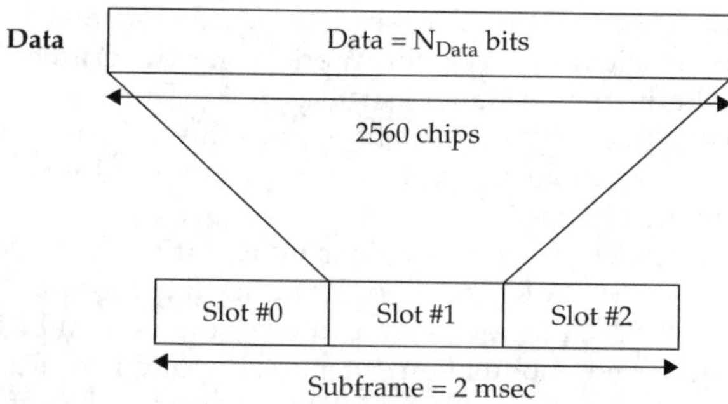

FIGURE 2-13 HS-DSCH time slot and subframe structure

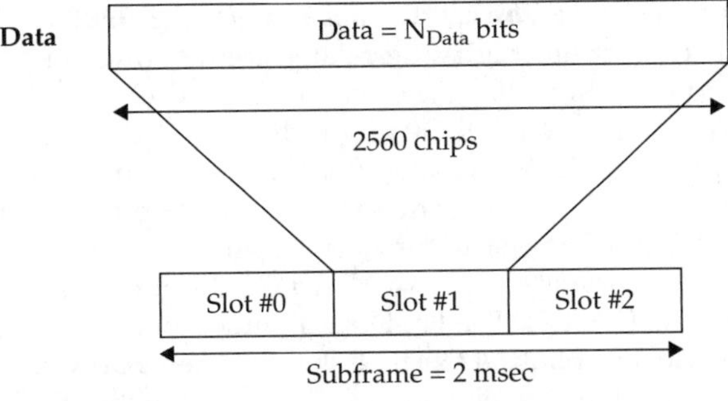

FIGURE 2-14 HS-SCCH time slot and subframe structure

for X multicodes at offset Y the codes are allocated contiguously ($C_{CH,16,Y} \ldots C_{CH,16,Y+X-1}$). This information is signaled over the HS-SCCH. This channel can be scrambled by either the primary code or a secondary code. The time slot and subframe structure is given in Figure 2-14.

The information carried in the HS-SCCH physical channel is used not only to alert the specific UE the data is intended for, but also to supply the UE receiver with some physical layer information so that it can prepare to demodulate the HS-DSCH channel.

The HS-SCCH will carry the downlink control signaling related to its associated HS-PDSCH transmission. Let us make a quick comment about the timing relationship between the P-CCPCH, HS-DSCH, and HS-SCCH physical channels, as shown in Figure 2-15.

There are five subframes per single 10 msec radio frame. The P-CCPCH and HS-SCCH channels are frame aligned. The HS-DSCH overlaps the HS-SCCH by a single time slot. This allows the UE to use the first two slots of the HS-SCCH subframe to set up the HS-DSCH demodulator. The one-slot overlap reduces the round-trip latency as well as the memory requirements of the UE.

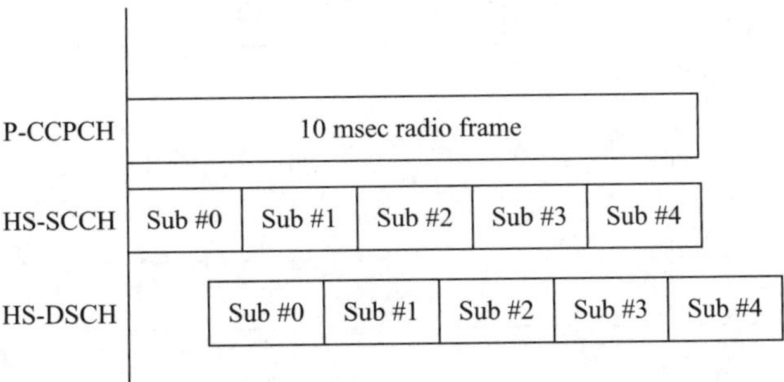

FIGURE 2-15 HS-DSCH and HS-SCCH subframe timing relationship

High Speed–Dedicated Physical Control Channel (HS-DPCCH) The HS-DPCCH carries uplink signaling related to the downlink HS-DSCH transmission. This signaling consists of Hybrid ARQ acknowledgment (HARQ-ACK) and Channel Quality Indication (CQI) and Precoding Control Interface (PCI). The ACK/NACK is carried in the first time slot, while the CQI/PCI is carried in the second and third time slots. The time slot and frame structure is given in Figure 2-16.

The SF is constant at 256. An HS-DPCCH can only exist together with an uplink DPCCH. The HS-DPCCH time slot timing relationship with respect to the uplink DPCH time slot is not necessarily time aligned. This depends on the downlink timing offset given to the DPCH.

These multicodes used on the HS-DSCH are dynamically allocated every TTI interval, which is fixed to 2 msec for HSDPA packet data transmissions. In comparing this HSDPA TTI value to that typically used for WCDMA CS voice applications, which is 20 msec for voice data and 40 msec for control data, the HSDPA value is significantly smaller. The fact that the scheduler was moved to the NodeB, as well as lower TTI intervals, significantly decreases the overall system delay and, as we will soon see, allows for accurate dynamic scheduling in order to track fast channel variations and thus more effectively use resources.

The dynamic scheduling is the strategy in which the NodeB determines how much of the available resources are allocated to the users waiting to be serviced. The wireless propagation conditions not only vary wildly over time, but also from one user's position in the cell to the next. The dynamic scheduling will utilize information fed back to the NodeB from the UE, providing an indication of the channel conditions "seen" by the UE. The NodeB scheduler will utilize this information from all UEs to determine the amount of the shared resources to distribute. This feedback information will essentially track the instantaneous channel conditions.

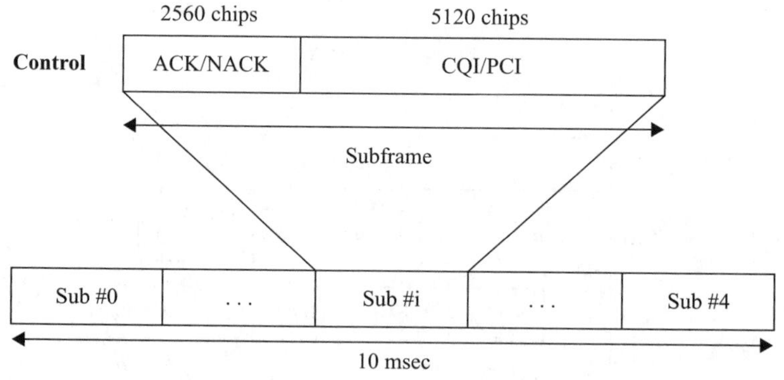

FIGURE 2-16 HS-DPCCH subframe structure

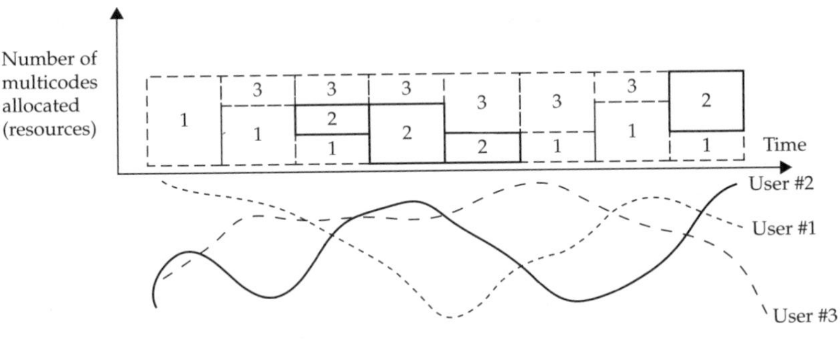

FIGURE 2-17 Three-user example of HSDPA scheduling

In Figure 2-17 we provide an example where three users (#1, #2, and #3) are sharing the HS-DSCH resources, which total 15 multicodes. We have arbitrarily chosen to allocate resources to the three users in multiples of 5 multicodes.

Here we see three varying channel profiles (or SNR) with the intention of pointing out three possibly different geographical locations within the NodeB coverage area (i.e., cell).

In addition to the number of multicodes the scheduler can assign to a UE, the scheduler can also vary the modulation scheme used. For HSDPA this corresponds to QPSK, 16-QAM, and 64-QAM. The use of Higher Order Modulation (HOM) provides for a more spectrally efficient transmission, especially when measured in bps/Hz. However, this benefit comes to us with a cost in that HOM requires a higher SNR in order to be effective. Hence users closer to the NodeB will more often enjoy HOM transmissions than those users at the cell edge. This is one argument in favor of femtocells.

HARQ is also another mechanism used by HSDPA to effectively deliver packet data transmission to users. Here a UE decodes each transport block and feeds back decoding status such as pass (via ACK) or fail (via NACK) to the NodeB. If the NodeB receives a pass from the UE, then the next transport block is transmitted. However, if the NodeB receives a fail from the UE, then the same transport block is retransmitted, this time with either more or different redundancy information. This pertains to either Incremental Redundancy or Chase Combining. The UE will use each retransmission to help it successfully decode the packet (transport lock) that was previously in error. Here the UE uses a previously obtained transport block to combine with the currently observed block to improve decoding performance. This physical layer retransmission protocol will continue for up to four times, after which the transport block will be flushed from the NodeB's buffers and the next transport block is transmitted. The RLC layer will then have the option to request further retransmissions if required.

HOM provides the dynamic scheduler with additional features. Recall HOM would typically sacrifice power efficiency for bandwidth efficiency. The benefit to the dynamic scheduler will be evident especially when a small set of channelization codes and large amount transmit power are available.

We have just described the benefits of dynamic scheduling and HARQ while maintaining low latency in the system. These requirements clearly indicate that their respective position in the overall network architecture must be closer to the radio interface. This was accomplished by the introduction of the MAC sublayer, called MAC-hs, which is responsible for dynamic scheduling, HARQ, etc., of the HSDPA services.

This very benefit imposes a new restriction. Since the MAC-hs function resides in the NodeB, it is not possible to transmit the same information from different NodeBs. Hence this prohibits the use of inter-NodeB soft handover. Ordinarily, this loss in macro-diversity would be alarming; however, the dynamic scheduler reduces this loss. Figure 2-18 shows a situation when the HSDPA services are provided by the serving cell and the dedicated services are provided by not only the serving cell but also the nonserving cell. It is worthwhile mentioning that any dedicated downlink channel can be in soft handover, though.

The UE sends feedback information to the NodeB called the Channel Quality Indication (CQI). The CQI is not mapped to a SNR; rather, it is an indication of transport block size the UE can currently accommodate. This would allow UEs with advanced receivers to differentiate themselves. The ACK/NACK information is also embedded in this uplink physical channel communication. In order to reduce latency due to the re-transmission, the UE should alert the NodeB of a Pass/Fail decision as soon as possible.

Radio Resource Management is still the responsibility of the RNC. In Figure 2-19 we provide the user-plane protocol architecture for HSDPA.

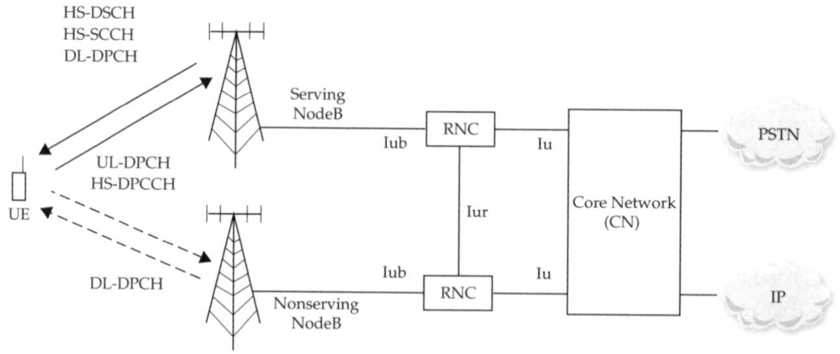

FIGURE 2-18 HSDPA network architecture

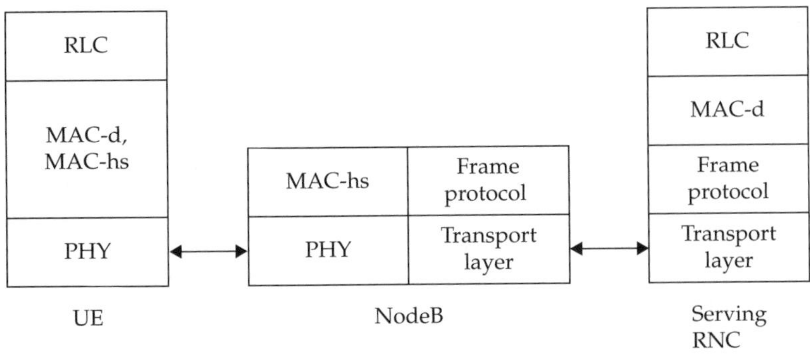

FIGURE 2-19 User-plane HSDPA protocol architecture

Release 6 introduced the FDPCH channel in order to accommodate the scenario when all the downlink traffic is carried over HS-DSCH; in this case only power control information is transmitted on the downlink dedicated channels.

Please note that even though soft handoff is not supported by HSDPA, hard handoffs do occur due to mobility. In this case, the UE will provide the measurement report to the RNC, while providing a listing of neighboring cells that satisfy certain system requirements. When one such cell is identified, the RNC will notify both the NodeB and UE of the update. When this occurs, the UE will flush its buffers and move on to the new cell. The NodeB will also flush packets in its buffers; indeed, this can lead to possible packet data loss in the UAM of the RRC. However, this can be minimized if the network properly plans and ensures packets are sent to the new cell in an orderly fashion.

Please note in the context of HSDPA services, the HARQ will control the PHY layer retransmission (up to a certain point) and will be used in addition to the RLC retransmissions if the AM RLC is enabled.

HSUPA Details

This enhancement addresses improving the uplink packet data capabilities by increasing the data rate, reducing latency, improving system capacity, etc., now with respect to the uplink. This enhancement was introduced into the 3GPP standard in the Release 6 version.

HSUPA also utilizes dynamic scheduling and Hybrid ARQ. Just like the downlink, the uplink has a TTI interval of 2 msec. However, the network can also use a 10 msec TTI if the overhead and system latency vary. The uplink transmission channel is called the Enhanced Dedicated Channel (E-DCH). This proposal was to go beyond the WCDMA Release 99 data rate of 384 Kbps in the uplink. Unlike HSDPA, which is a shared channel, HSUPA is a dedicated channel solution.

Here each HSUPA-capable UE has its own E-DCH that it wishes to transmit.

In the uplink, users don't need to share channelization codes. HOM on the uplink was introduced to the 3GPP standard in the Release 7 version as a way to further increase the uplink data rates.

The dynamic scheduler is located in the NodeB and indirectly controls the uplink data rate. This is accomplished by allocating varying levels of UE transmission power. Please recall that for HSUPA, just like WCDMA, the uplink is nonorthogonal. Hence more users transmitting or users transmitting with higher power leads to an increase in the uplink interference. This interference is both intracell as well as intercell.

The basic HSUPA principles are based on the NodeB sending grants to the UE and the UE sending requests to the NodeB for resources. Due to the nature of the uplink, the dynamic scheduler will typically schedule many users in parallel. In soft handoff, the serving cell has the responsibility for scheduling.

HARQ is also used in the uplink, whereas each transport block is transmitted by the UE, there is an ACK/NACK response from the NodeB. In soft handoff a few NodeBs will receive the UE-transmitted packet; as long as the UE gets a single ACK, the data was successfully transmitted to at least one NodeB. There can be only one E-DCH transport channel in the UE, whereas multiple DCHs can be multiplexed together.

An example of the HSUPA network architecture is provided in Figure 2-20. For similar reasons as with HSDPA, the dynamic scheduling and HARQ functions are located in the NodeB. This is shown in the figure. For HSUPA we have introduced relative grants, absolute grants and an indication of reception.

A new MAC entity, MAC-e, is introduced to both the UE and the NodeB. In the UE, the MAC-e is responsible for determining the

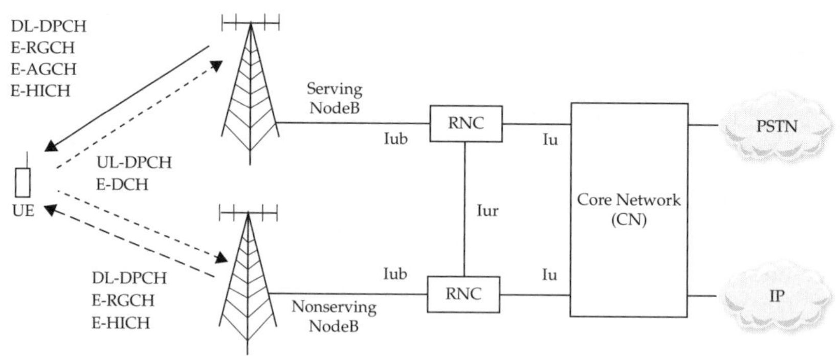

FIGURE 2-20 HSUPA network architecture

allowable data rates as provided by the scheduler. In other words, it is responsible for dynamic scheduling and HARQ operations.

To guarantee sequential delivery of packet data to the RLC protocol, a re-ordering feature is necessary in the RNC. This feature requires a new entity called the MAC-es. When the HSUPA experiences a soft handover, the HARQ retransmissions will have each piece of data delivered out of order and need to be compensated in the RNC. The MAC-es is responsible for correcting this behavior. The additional MAC entities are shown in Figure 2-21.

As discussed previously, the HARQ procedure requires knowledge of successful and nonsuccessful transmissions. This is accomplished with the assistance of ACK/NACK indicators. This information is in the E-DCH Hybrid ARQ Indicator Channel (E-HICH).

Recall, grants are issued to the UE by the NodeB dynamic scheduler. This information is in the shared E-DCH channel called E-DCH Absolute Grant Channel (E-AGCH). The E-AGCH is transmitted only by the serving cell. Moreover, grants can also be sent to the UE via E-DCH Relative Grant Channel (E-RGCH) transmissions.

E-RGCH is typically used for finer adjustments, and E-AGCH is typically used for large adjustments in the data rate.

The NodeB requires information about the packet data; this is in the form of a Control Channel and is presented as the E-DCH Dedicated Physical Control Channel (E-DPCCH).

Grants are expressed as E-DPDCH/DPCCH power ratios. They are updated by the NodeB by transmitting an absolute grant or a relative grant. The E-AGCH channel is used to make absolute changes to the serving grant transmitted from serving cell only. The E-RGCH channel is used to make relative changes to the serving grant. E-RGCH from serving cell has three values: UP, DOWN, and HOLD. The UP command tells the UE to increase the allowed E-DPDCH/DPCCH power ratio. For nonserving cells it has two values: DTX and DOWN.

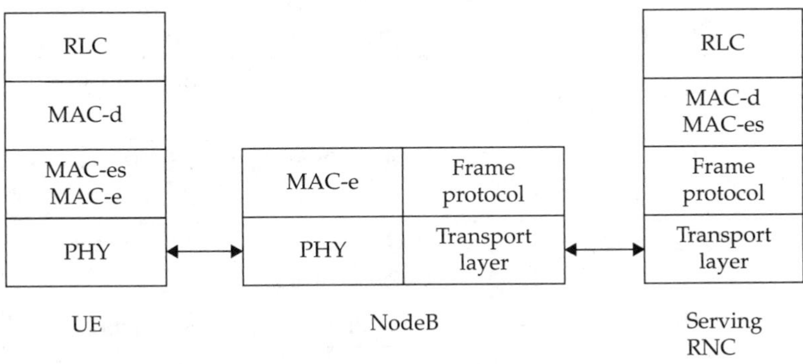

Figure 2-21 User-plane HSUPA protocol architecture

The E-TFC is responsible for determining the transport format of the E-DCH. Note that the dynamic scheduler allocates resources between UEs, while the E-TFC allocates resources within the UE data flow.

Let's take this time to provide a simple comparison of how the HSPA functionality is partitioned across the network. For HSDPA, Dynamic Scheduling and buffering are located in the NodeB, while the reordering is in the UE. For HUSPA, Dynamic Scheduling is located in the NodeB, buffering is in the UE, and reordering is in the RNC. The E-DPCCH carries information that helps decode the E-DPDCH:

- **Happy Bit** Is used to indicate if the UE can benefit by a higher serving grant (i.e., a higher power allocation).
- **E-TFCI** E-DCH refers to the transport format combination indicator.
- **RSN** A two-bit retransmission sequence number to keep track of retransmissions and support in-sequence delivery to the RLC.

The uplink and downlink can transmit DPCCH and DPDCH at the same time as the E-DPCCH and E-DPDCH; however, priority is given to the DPCH. The remaining power is provided to the E-DCH transmission. Based on the downlink grants, available transmit power; and available data in buffer, the UE is responsible for selecting a transport block size to use. The E-DCH can be mapped to one or several E-DPDCHs. The possible data rates are given in Table 2-2.

Next we provide an overview of the physical channels used in supporting the HSUPA services.

E-DCH Category	E-DPDCHs	Data Rates	
		10 msec	2 msec
1	1 × SF4	0.7 Mbps	–
2	2 × SF4	1.4 Mbps	1.92 Mbps
3	2 × SF4	1.4 Mbps	–
4	2 × SF2	2 Mbps	3.84 Mbps
5	2 × SF2	2 Mbps	–
6	2 × SF4 + 2 × SF2	2 Mbps	5.76 Mbps

TABLE 2-2 E-DCH Categories for the UE

E-DCH Hybrid ARQ Indicator Channel (E-HICH) This is a downlink dedicated physical channel providing HARQ information to the UE regarding the decoding of the E-DCH. The E-HICH will transmit either an ACK or a NACK if the cell is a serving cell; otherwise, only an ACK is transmitted. This is to reduce the downlink transmit power and interference. The E-HICH transmits a single bit of information for a duration of either 2 msec or 8 msec. The SF used is constant and set to 128. The E-HICH and E-RGCH use the same channelization code and scrambling code. In a softer handoff, the UE will soft-combine the E-HICH from each cell. The E-HICH and E-RGCH use 40-bit orthogonal sequences to multiplex multiple E-HICHs and E-RGCHs on a single downlink code.

E-DCH Absolute Grant Channel (E-AGCH) This is a downlink shared channel providing absolute scheduling grants that conveys the maximum E-DPDCH/DPCCH power ratio and activation flag to activate or de-activate a particular HARQ process. The spreading factor used is constant and set to 256. R = 1/3 convolutional coding is used, creating 2 msec and repeated 5× to 10 msec time duration.

E-DCH Relative Grant Channel (E-RGCH) This is a downlink dedicated physical channel providing relative grants at a rate of one per TTI. Values allowed are UP, HOLD, and DOWN (for the serving cell). It has a time duration of 2 msec or 8 msec. For nonserving cells, the values are HOLD and DOWN and for this case the relative grant has a time duration of 10 msec. This means the three time slots (2 msec time duration) are repeated five times. The RRC signaling informs the UE which signature (orthogonal sequence) the E-RGCH is using.

E-DCH Dedicated Physical Control Channel (E-DPCCH) This is an uplink dedicated control channel transmitting RSN, E-TFCI, and Happy Bit (Rate Request). The SF used is 256. The E-DPCCH is only transmitted when the E-DPDCH is transmitted. The E-DPCCH timing is aligned with that of the DPCCH timing.

In this section we will discuss the HSUPA relevant channels to support uplink packet access. A major difference is that HSUPA is not a shared channel but a dedicated channel. This implies it operates in soft handoff scenarios. The reason for this is as follows: On the downlink the NodeB resources can be controlled and allocated to a single UE at a time (if needed). However, the uplink resources (UE transmissions) cannot be shared and thus act very much like the DPCH signals used for WCDMA.

The uplink restriction is that when E-DCH is used, the maximum DCH data rate is 64 KBps. The E-DPDCH supports simultaneous transmission of two SF = 2 codes and two SF = 4 codes, which leads to a maximum physical-layer bit rate of 5.76 MBps. This was shown in Table 2-2.

The E-DPDCH channel requires the uplink DPCCH to be simultaneously transmitted to aid in channel estimation, SIR estimation, and power control. Also the E-DPDCH requires transmission of the E-DPCCH to disclose the E-DPDCH format to the NodeB. The uplink timing relationship of the E-DCH and DPCH channels is time slot aligned.

E-DCH HARQ Indicator Channel (E-HICH) E-HICH has a fixed SF = 128 used to carry E-DCH HARQ ACK indicators. The indicator is transmitted using 3 or 12 consecutive slots, which are used with the E-DCH TTI set to 2 msec and 10 msec. Each slot contains 40 bits, which are set to an orthogonal sequence. The time slot and frame structure is given in Figure 2-22.

The HARQ indicator is allowed to take on ACK or NACK values for radio link sets containing the serving E-DCH. For those cases when the radio link does not contain the serving E-DCH, the HARQ indicators can only take on ACK or DTX values. This latter case was created in order to reduce the downlink transmission power. Hence the UE will continue to retransmit until at least one cell responds with an ACK indicator. The relative timing of this physical channel will be discussed later in this section. No channel coding is applied to this channel.

E-DCH Relative Grant Channel (E-RGCH) The E-RGCH has a constant SF = 128 and is used to carry the uplink E-DCH Relative Grants that are transmitted using 3, 12, or 15 consecutive time slots. The time slot and frame structure is shown in Figure 2-23.

Radio links in the serving E-DCH cell use the 3- and 12-slot duration for an E-DCH TTI of 2 msec and 10 msec, respectively. The 15-slot duration is used for radio links not in the serving E-DCH cell. A 40-bit orthogonal sequence is transmitted in each time slot. The E-RGCH is used to signal to the UE a relative power up or down command to control the E-DPDCH transmission power. No channel coding is applied to this channel.

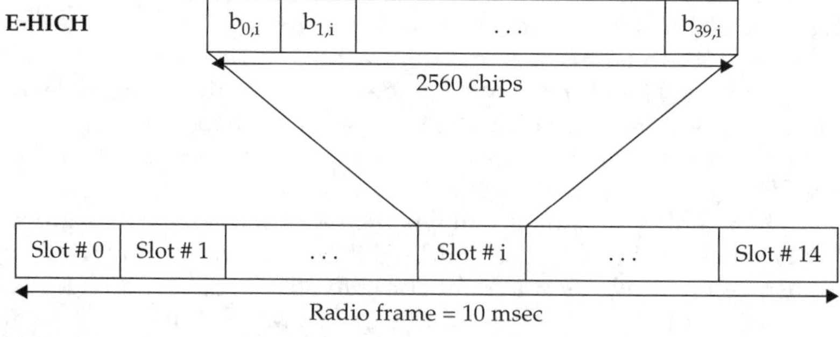

Figure 2-22 E-HICH time slot and frame structure

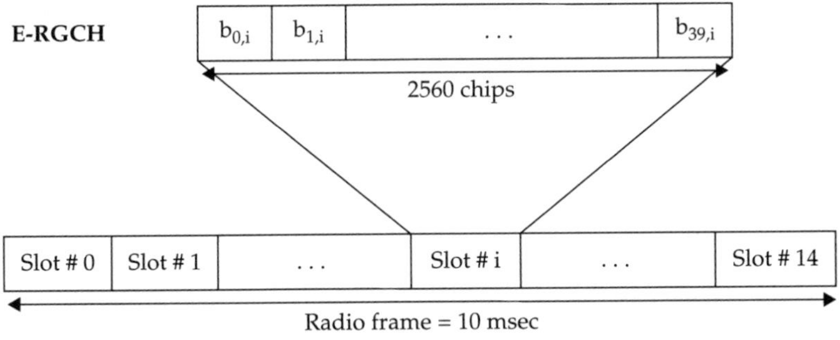

FIGURE 2-23 E-RGCH time slot and frame structure

E-DCH Absolute Grant Channel (E-AGCH) The E-AGCH has a constant SF = 256 and is used to carry the E-DCH absolute Grant. The time slot and frame structure is given in Figure 2-24.

The absolute grant is used to tell the UE the maximum relative transmission power it is allowed to use. This relative power is E-DPDCH with respect to DPCCH. This channel contains a 16-bit CRC and an R = 1/3 convolutional code and follows by rate matching. The resulting bit stream consists of 60 bits, which are transmitted over three consecutive time slots.

E-DCH The E-DCH consists of E-DPCCH and E-DPDCH channels. The E-DPDCH will carry the E-DCH transport channel, whereas the E-DPCCH will carry the control information. For the most part, the E-DPDCH and E-DPCCH are transmitted simultaneously. The E-DPCCH will not be transmitted unless the DPCCH is also transmitted. The time slot and frame structure are provided in Figure 2-25.

The SF on E-DPDCH can vary from 2 up to 256, whereas the SF on E-DPCCH is constant at 256. The Channelization code for E-DPCCH

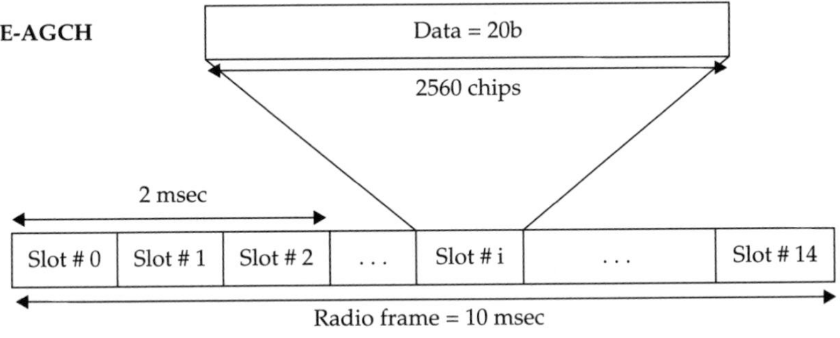

FIGURE 2-24 E-AGCH time slot and frame structure

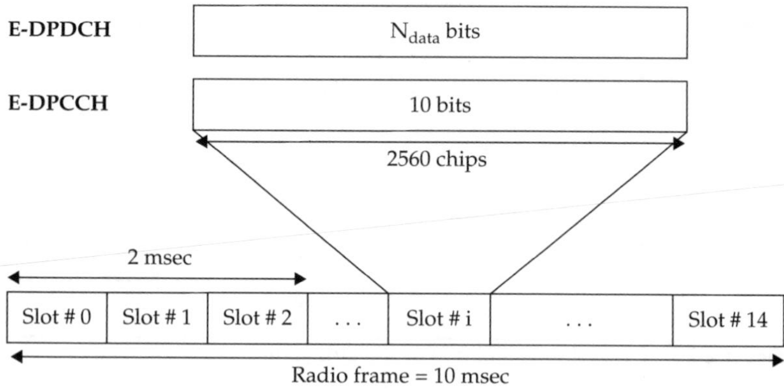

FIGURE 2-25 E-DCH time slot and frame structure

is always $C_{CH,256,1}$. The HARQ process is similar to that presented earlier when discussing HSDPA. A rate R = 1/3 Turbo Coding FEC is used. A point worth mentioning here is that unlike HSDPA, HSUPA has two TTI time durations of 2 msec and 10 msec. The 2 msec TTI has the benefit of shorter delays, while the 10 msec TTI can have improved performance at the cell boundary.

Network Architecture

In this subsection we will provide the network architecture of the combined functionality of HSDPA and HSUPA, generally called HSPA. We will also show the migration in their functionality, specifically the pattern of moving functionality closer to the physical (radio access) layer interface. First we provide the Release 99 Network Architecture, as shown in Figure 2-26.

Here we defined and differentiated RNCs. S-RNC is the Serving RNC, which is connected to the core network, while D-RNC is the Drift RNC, which handles the call to the NodeB. It is extremely visible that the RNC maintains control over the resources.

Next we provide the HSPA network architecture, shown in Figure 2-27. Note this requires a change in the overall RRM architecture. It is fairly easy to note that the functionality partitioning has changed and, in fact, moved closer to the physical channel signal processing or NodeB. The tremendous benefit gained is a reduction in latency, since signaling across components necessary for closed-loop functions have been moved into a single component.

In an effort to compare the WCDMA, HSDPA, and HSUPA features, Table 2-3 provides a brief feature comparison table for the DCH-, HS-DSCH-, and E-DCH-related channels.

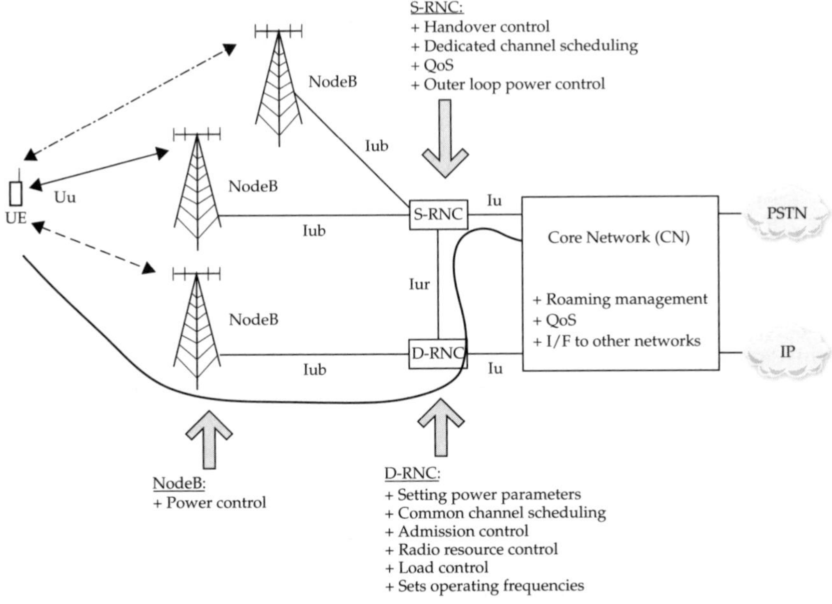

Figure 2-26 WCDMA network architecture functionality partitioning

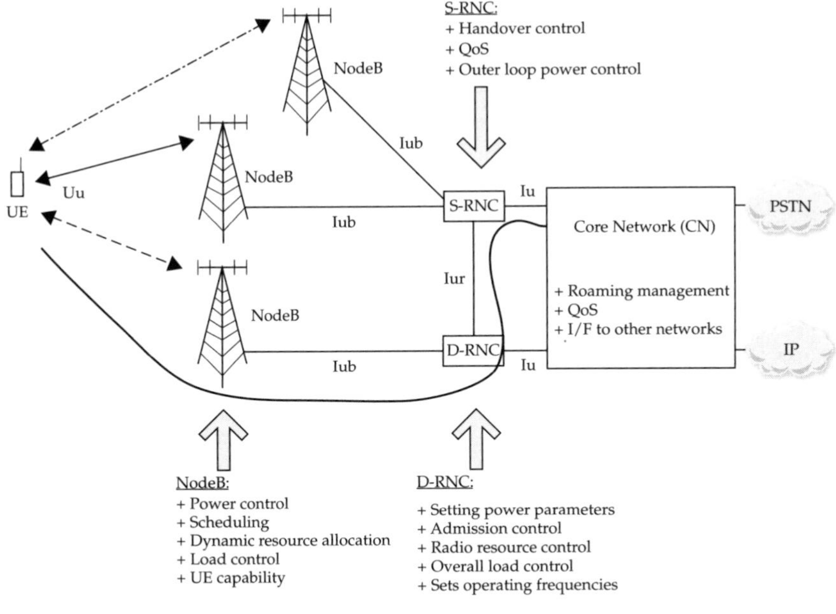

Figure 2-27 HSPA network architecture functionality partitioning

Feature	DCH	HSDPA	HSUPA
Variable SF	Yes	No	Yes
Fast power control	Yes	No	Yes
Adaptive modulation	No	Yes	Yes
TTI (msec)	10, 20, 40, 80	2	2, 10
SHO	Yes	No	Yes
NodeB scheduling	No	Yes	Yes
HARQ	No	Yes	Yes
Shared/dedicated	Dedicated	Shared	Dedicated

Table 2-3 WCDMA and HSPA Feature Comparison

Performance Results

In this subsection we provide some simulation results for HSDPA, as shown in Figure 2-28. The following are simulation results comparing the Type 3 to the Type 3i receivers, assuming perfect knowledge of the interfering channel matrix. Here we see for large geometries 16-QAM offers significant throughput advantages over QPSK. Also as the geometry decreases, a crossover region exists such that QPSK

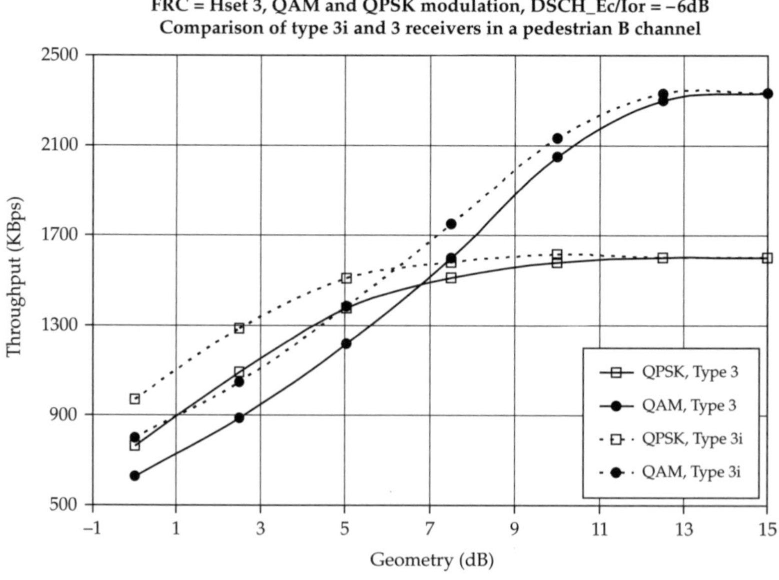

Figure 2-28 QPSK and 16-QAM performance comparison in a PB3 channel

performs better than 16-QAM. Finally, we can see that interference-aware receivers perform better at the lower geometries, where the UE is closer to the cell edge and has increased visibility to interference from other cells.

The geometry is defined as $G = I_{or}/I_{oc}$, where I_{or} is the received signal power and I_{oc} is the other cell's interference power.

2.1.4 HSPA Evolution

As discussed previously, the 3GPP standard Release 5 version introduced HSDPA and Release 6 introduced HSUPA, thus providing a very good solution to packet data applications not only for downlink, but also for uplink. Next we discuss features added in Release 7 that aim to further improve and evolve the overall HSPA user experience.

Multiple Input Multiple Output (MIMO)

MIMO is a technique used to improve system performance that utilizes multiple transmit and receive antennas. Among other benefits, we will focus on increasing the user's data rate by transmitting parallel data streams. The MIMO systems use multiple transmit antennas in order to effectively create parallel spatial streams to be transmitted. The initial introduction into the 3GPP standard provides both single and dual spatial streams using two transmit antennas. For the single spatial stream case, the precoding weights are those used in the Release 99 version, specifically for Closed Loop Transmit Diversity (CLTD). Here we show the two weights (for antennas #1 and #2) used for the transmit diversity case for either WCDMA or HSDPA applications:

$$w_1 = \frac{1}{\sqrt{2}} \quad \text{and} \quad w_2 = \frac{1}{\sqrt{2}}\{1 + j1, 1 - j1, -1 + j1, -1 - j1\}$$

For the dual spatial stream case, the precoding weights can be constructed in matrix notation. The weights of the second stream (i.e., w_3, w_4) are orthogonal to the first stream and are chosen such that the dual streams are uncorrelated with each other. The details on the choice of the antenna weights are left to the reader to investigate in line with the 3GPP standard [2]. In Figure 2-29 we provide a simplified block diagram of the MIMO technique applied to the HSDPA transmission, assuming dual-transmit antennas are used.

A point worthy of mention regarding MIMO system performance is the following: The additional complexity of a MIMO system over a SISO system is tremendous. This includes an increase in functionality in not only the NodeB, but also the UE. Hence the performance benefits should outweigh this necessary burden of complexity. When characterizing the performance of the 3GPP-based MIMO system, one can determine the performance improvement occurs more in micro/

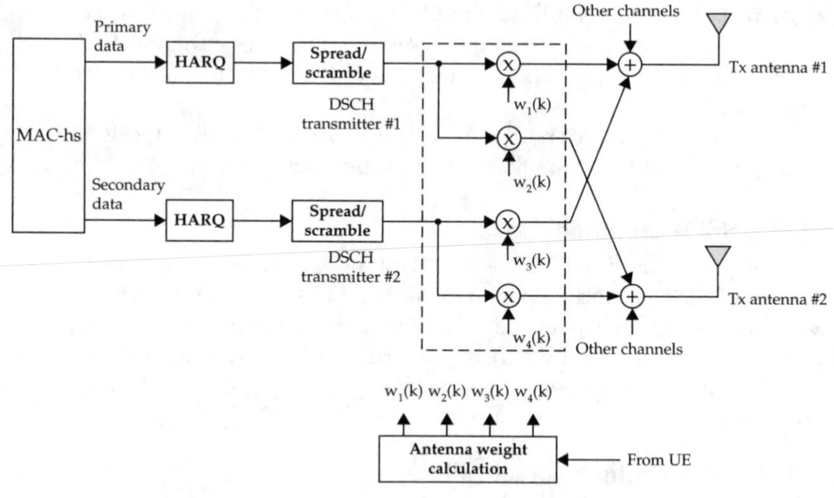

FIGURE 2-29 HSDPA MIMO downlink diagram

pico/femto cellular environments than in macro environments, not to mention lower-mobility mobiles. Hence when considering performance improvement such as, but not limited to, MIMO, clear requirements should be a prerequisite to the system design and deployment.

Every TTI, the HS-SCCH will inform the UE of the number of channelization codes, modulation scheme, and precoding matrix, etc. ACK and NACK information per stream is required at the NodeB to support the dynamic scheduling.

The dedicated uplink channel HS-DPCCH transmits the ACK/NACK, PCI, and CQI information to the NodeB. For the dual-stream case, the ACK/NACK for each stream is jointly encoded. The precoding control information (PCI) informs the NodeB about which precoding matrix is the best for the UE.

MIMO support is not required for all UEs. Please note as part of the process where the UE informs the NodeB about its capabilities and category, the NodeB will have knowledge about that particular UE's limitations.

Higher-Order Modulation (HOM)

As previously discussed, HOM sacrifices power efficiency in order to obtain bandwidth efficiency. Hence if the propagation conditions permit, then the use of HOM can be an attractive solution. Release 7 introduces 64-QAM on downlink and 16-QAM on the uplink to address these particular circumstances.

In order to provide the reader with a high-level appreciation of the wonderful opportunities that MIMO and HOM afford us, we have created Table 2-4 exposing the maximum possible data rates. At this particular point, the reader can see how HSPA evolution offers

DL Peak Data Rates			UL Peak Data Rates	
16-QAM	64-QAM	64-QAM+MIMO	BPSK/QPSK	16-QAM
14 Mbps	21 Mbps	42 Mbps	5.7 Mbps	11 Mbps

Table 2-4 Maximum Uplink and Downlink Data Rates

a very attractive solution and road map to support packet data networks and services.

Continuous Packet Connectivity (CPC)

We have presented the importance of the 3GPP network to support packet data communications. This support involves adding packet access capabilities to both the uplink and downlink directions.

This next technique is targeted to reduce overhead for services that require maintaining a communication link but that don't always have a constant flow of data to transmit. An example of such a service is VoIP. If one were to carefully review the packet data traffic statistics, one would learn it is bursty in nature. In order to make the users' experience more pleasant, it is desirable to have the HSDPA and HSUPA channels already set up in order to support the transmission and reception of packet data.

However, having these channels configured and not transmitting any packet data is detrimental to both the network and to the user. For example, having the UE transmit a dedicated control channel when there is no data to transmit is highly inefficient and wasteful. This causes unnecessary interference in the uplink (adding to rise over thermal noise), not to mention considerable drain on the UE battery (increasing power consumption and decreasing UE talk time).

Similarly, we can apply the same philosophy to the receiver. Having the UE constantly monitor the downlink control channels is not the most efficient technique, especially when one is aiming to reduce the power consumption of the UE.

On the other hand, completely removing the requirement of having the UE receive and transmit HSDPA and HSUPA, respectively, will only complicate matters and produce inefficient setup and tear-down overhead in the overall system. Hence it would be desirable to stay in the CELL_DCH RRC state, but benefit from the topics already discussed.

Continuous Packet Connectivity was introduced in the 3GPP standard in Release 7; it primarily consists of three features:

- **Discontinuous Reception (DRX)** Allows the UE to periodically monitor the downlink channels in order to save power consumption. This avoids the UE having to always have its receiver powered on to receive a signal that is not always intended for it.

- **Discontinuous Transmission (DTX)** Allows the UE to periodically transmit the uplink channels in order to reduce the uplink interference and reduce the UE power consumption. In this situation, the UE will only transmit when it has information to transmit.
- **HS-SCCH–Less** Reduces HSDPA overhead for low–data rate applications, such as VoIP. The need to carry out the first step of demodulating the control channel is removed to reduce latency and overhead.

These features will give the user the appearance of an "always on" connection, while providing overall system benefits.

In the DTX case, the UE generates interference since the uplink DPCCH channel is transmitted as long as E-DCH channels are configured by the higher layers. One can completely turn off the uplink transmissions; however, this has a detrimental effect on the uplink synchronization, power control, etc. Hence having a UE periodically perform DTX will create a tremendous benefit for all these issues. The UE DTX cycles are configured in the UE and NodeB by the RNC.

When the UE is DTX-ing, the NodeB cannot measure uplink SIR and hence has nothing to transmit as power control commands as far as the downlink is concerned (via FDPCH). In order to help the NodeB reception during these on/off transition time durations, preambles and postambles are used by the UE transmission for improved system performance.

When DTX cycle is enabled, the ACK/NACK commands are transmitted regardless of DTX cycles. However, uplink CQI reporting has a lower priority than the DTX cycle, and thus these signals are only transmitted if they are time-aligned with the UE transmission bursts.

In the DRX case, the UE should monitor the HS-SCCH, E-AGCH, and E-RGCH during the UE DRX cycle. Just as the case for HSDPA, the UE will monitor the ACK/NACK on the E-HICH. The DTX and DRX cycles are configured and activated by RRC signaling.

The final CPC-related topic addresses the HS-SCCH–less operation. Recall how in Release 6 the F-DPCH was introduced to save channelization code space. Another potential savings can be had by reducing the HS-SCCH usage. Here the UE will decode the HS-DSCH without the assistance of HS-SCCH. This involves the network configuring some predefined set of transport formats and having the UE perform blind detection. This will easily prove to be useful in the VoIP application.

At this point, we have a physical layer that supports packet access. We have introduced some improvements to the uplink and downlink in order to address the burstiness of the packet traffic, and we have simplified decoding of small-sized packets.

Enhanced CELL_FACH

Recall the preceding modifications were actually techniques that the network can use to extend the time duration a UE would be in the CELL_DCH RRC state while offering improved overall system performance.

A point to make here is if for a prolonged period of time, there is no transmission due to the burstiness of the packet traffic, the UE will be moved from the CELL_DCH to either the CELL_FACH or CELL_PCH RRC state. For the CELL_FACH RRC state, the FACH channel is required to change the RRC state back to the CELL_DCH. This incurs a potentially large and inefficient latency.

Release 7 allows the HS-DSCH to be used also in the CELL_FACH RRC state. This is called Enhanced CELL_FACH. When HS-DSCH is used for this scenario, there is no uplink to signal CQI, or ACK/NACK information. For completion, we will mention that the CELL_PCH information can also be transmitted over the HS-DSCH. Hence the UE can demodulate the HS-DSCH after it had received the paging indication. This is all part of the migration path to packet-deliver services through the HS-DSCH channel.

Advanced Receivers

Besides network latency reduction, decreased setup times, etc., the user and network system performance can be further improved with the use of advanced receivers. The philosophy of the 3GPP standards body has been to agree on a reference receiver structure so that delegates can submit performance contributions. The initial phase of the submittals is to align the simulation assumptions and to agree on a reference receiver structure. The last phase of setting the performance requirements is for delegates to submit performance results that contain an implementation margin. By this we mean enough room for implementation degradation (when compared to the simulation results submitted during the first phase). Historically this has been anywhere from 1 dB to 3 dB, but this value is left up to each individual contribution.

The benefit of such a phased approach is that it aligns every company's simulation assumptions and physical-layer details at the beginning of the performance requirements stage. This initial alignment is the opportune time to find performance and system-related issues. This phased approach also allows contributions to be made based on proprietary algorithms and structures if need be. The final results are typically averaged and used for the standard requirements. Table 2-5 shows the defined types and receiver structures that have been incorporated in the 3GPP standards body.

The receivers shown in the table improve performance by increasing peak data rates across the cell region. They also increase system capacity, increase cell size, increase mitigation to rise in interference, etc. There are many forms of improvement; they depend on the deployment scenario and link budget used in the system capacity phase.

Type	Reference Receiver Structure
0	RAKE
1	Spatial Receive Diversity (M = 2 antennas)
2	Equalizer (fractionally spaced)
3	Spatial Diversity + Equalizer
2i	Equalizer with Interference Suppression
3i	Spatial Diversity + Equalizer with Interference Suppression
M	MIMO

TABLE 2-5 3GPP Reference Receivers

2.1.5 Long-Term Evolution (LTE) Overview

In order to evolve the packet capabilities of the WCDMA/HSPA systems, 3GPP is supporting the LTE radio access technology. We will review the expected benefits of this evolved system. Similarly, the core network will also evolve, and this improvement is known as System Architecture Evolution (SAE). These new requirements are provided in the Release 8 version of the 3GPP 36.XXX series of standards [7–12]. The next-generation networks (NGN) are moving to an all-IP-based network; LTE and SAE will support the NGN using IP Multimedia Subsystem (IMS).

LTE will use some of the same principles as HSPA, specifically regarding Scheduling, Shared Data Channels, HARQ, etc. LTE had certain objectives in mind while the standard was created. They target the following categories:

- Data rate requirements. The downlink and uplink peak data rate requirements are 100 Mbps and 50 Mbps, respectively. This is assuming a 20 MHz bandwidth of spectrum.
- Support for both FDD and TDD systems.
- Support for mobility.
- Reduced latency.
- Improved user experience.
- Coexistence with legacy 3GPP systems.
- Flexible spectrum deployment using various BW configurations.
- Flexible spectrum deployment using a variety of frequency bands.
- Improved coverage.
- Increased security.
- Increased system capacity.

In order to achieve these system performance targets, the physical and network architectures were modified. An example of this begins with a change in the multiple access technique. On the downlink, Orthogonal Frequency Division Multiplexing (OFDM) is chosen for transmission; it comes with the following benefits:

- OFDM offers flexible allocation of bandwidth.
- OFDM has been successfully deployed in single-frequency networks (SFNs) utilizing the cyclic prefix (CP) to help combat frequency selective fading (FSF).
- OFDM also expands the capability of the scheduler by including the frequency domain. The HSDPA scheduler made use of time and channelization codes, while in LTE it will make use of time and frequency channels.

However, on the uplink the UE complexity is an essential obstacle that deserves special attention. It is for this reason that single-carrier frequency division multiple access (SC-FDMA) is chosen for transmission and comes with the following benefit:

- SC-FDMA has a lower peak-to-average power ratio (PAPR) than the multicarrier OFDM.

In either the uplink or downlink case the scheduling of the time-frequency response takes place. The scheduling is updated every 1 msec or every subframe. HARQ with retransmissions will also be used in LTE to improve performance.

In contrast to WCDMA and HSDPA requirements, LTE will begin with a reference receiver that assumes dual spatial diversity capability. This is expected to offer tremendous performance gains that can be used across a multitude of avenues. Also, in LTE further improved performance will also be achieved using multiple antennas at the transmitter and receiver (i.e., MIMO application) along with the use of HOM.

Uplink and Downlink Channels

In the next subsections we will explore the downlink and uplink channel structures for the LTE physical interface. There are two frame type structures: Type 1 is used for FDD, and Type 2 is used for TDD. Moreover, the support of both full duplex and half duplex exists. We will focus on the FDD aspect of the standard.

Downlink Channel Structures Some specifics about the physical layer lead us to start with the following definitions. A radio frame has a time duration of 10 msec. It consists of ten subframes, each of size 1 msec. Each subframe has two time slots, where each slot has a time duration of 0.5 msec. You can see this structure in Figure 2-30.

The downlink transmission uses OFDM. One resource element is an OFDM subcarrier with the subcarrier spacing of Δf = 15 KHz.

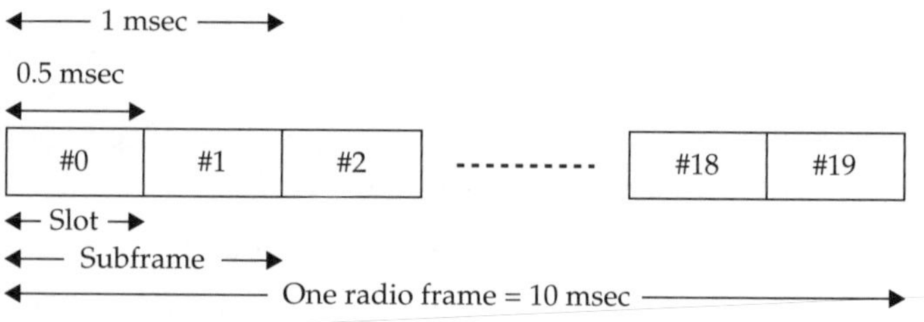

FIGURE 2-30 LTE downlink radio frame structure

A resource block consists of 12 consecutive subcarriers, resulting in a resource block bandwidth of 12 * 15K = 180 KHz. There is also a Δf = 7.5 KHz, but this is used only in an MBMS dedicated cell. The resource block is shown in Figure 2-31, defined by the number of subcarriers ($N_{subcarriers}$) and number of symbols ($N_{symbols}$). Another parameter that is also available to the LTE physical layer is the number of resource blocks ($N_{resource\text{-}blocks}$).

Table 2-6 shows what resource block parameters change when we change the Cyclic Prefix modes.

A symbol time duration is defined as T = 1 / Δf ~ 66.7 μsec. As typically performed in OFDM-based communications, a Cyclic Prefix

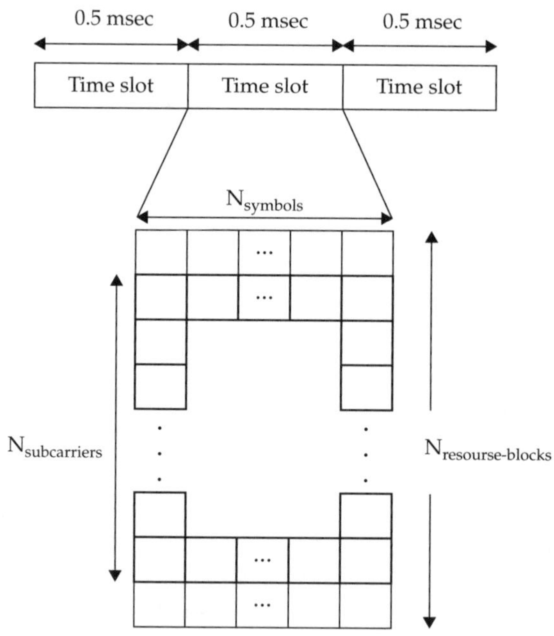

FIGURE 2-31 Downlink resource block definition

Modes of Operation	Δf	$N_{subcarriers}$	$N_{symbols}$
Normal CP	15 KHz	12	7
Extended CP	15 KHz	12	6
Extended CP	7.5 KHz	24	3

TABLE 2-6 Downlink Physical Resource Block Parameters

(CP) is inserted in order to utilize the cyclic property. As shown in Table 2-6, there are two CP lengths: a normal CP and an extended CP. In using the extended CP, the $T_{cp} = 1 / (4 * \Delta f) = 16.7$ µsec; moreover, when considering $T_{slot} = 0.5$ msec there are six OFDM symbols. During one time slot, there are seven OFDM symbols or 12 * 7 = 84 resource elements.

A downlink reference signal is inserted into the symbol stream to aid the UE in coherent demodulation. In order to most efficiently estimate the channel impulse response, these pilot (or reference) symbols are time multiplexed and frequency domain interlaced. In the downlink, scrambling is applied to all transport channels. The forward error correction (FEC) technique used is a R = 1/3 Turbo code. The downlink supports QPSK, 16-QAM, 64-QAM for DL-SCH. The downlink scheduling is carried out at a subframes rate of 1 msec. MIMO is supported with either two or four transmit antennas and either two or four receive antennas. As a side note, QPSK modulation is used for all control channels.

Uplink Channel Structures The uplink transmission technique is called Single Carrier Frequency Division Multiple Access (SC-FDMA), which offers resource assignment flexibility as well as orthogonality for multiple access, both of which are important for this system. The SC-FDMA transmission block diagram is shown in Figure 2-32, where we have shown the uplink data channel multiplexed with the control channel.

The Discrete Fourier Transform (DFT) size indicates the transmitted signal BW. The frequency mapping indicates which

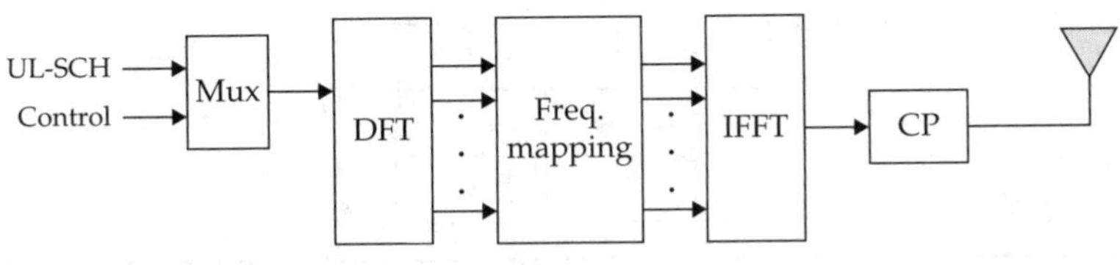

FIGURE 2-32 LTE uplink SC-FDMA signal generation

Modes of Operation	Δf	$N_{subcarriers}$	$N_{symbols}$
Normal CP	15 KHz	12	7
Extended CP	15 KHz	12	6

TABLE 2-7 Uplink Physical Resource Block Parameters

occupied frequency to use for transmission, and the Cyclic Prefix (CP) is inserted to allow frequency domain equalization. A resource block organization similar to the downlink is used in the uplink and is shown in Figure 2-33.

The concept of a 10 msec radio frame, a 1 msec subframe, and 0.5 msec time slot all applies to the uplink. The uplink resource blocks assigned to an UE must be adjacent to each other in the frequency domain. The uplink reference signals are time multiplexed. The uplink control signaling consists of HARQ ACK/NACKs, CQI reports, etc. We will discuss these channels in more detail in the following sections. The uplink supports QPSK, 16-QAM, and 64-QAM (optional) for UL-SCH.

Table 2-7 shows what resource block parameters change when we change the Cyclic Prefix modes.

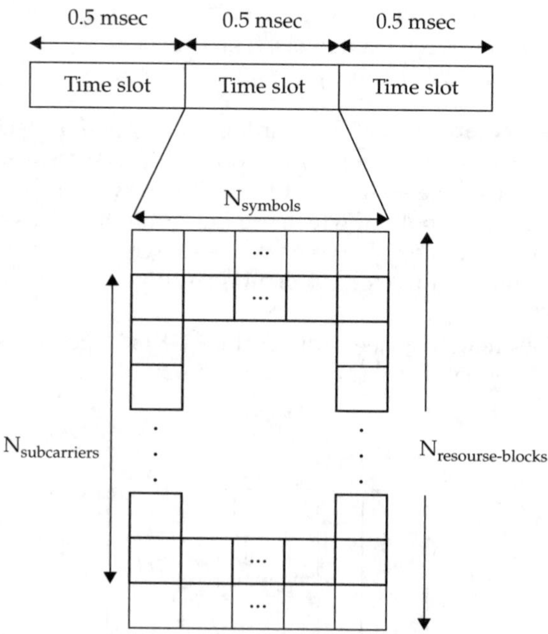

FIGURE 2-33 Uplink resource block definition

Network Architecture

In this subsection we present the network architecture specified for LTE, and whenever possible we compare the differences from the HSPA network. We provide a description for each of the protocol layers to be mentioned and shown in Figure 2-34. The Packet Data Convergence Protocol (PDCP) handles IP header compression, ciphering, etc. This is located in the eNodeB, where we have used the notation enhanced-NodeB to differentiate between the earlier versions of this component. The Radio Link Control (RLC) handles segmentation, retransmission, in-sequence delivery to higher layers, etc. This is now located in the eNodeB. The Medium Access Control (MAC) handles HARQ re-transmission, scheduling, logical channel multiplexing, etc. Finally, the Physical (PHY) layer handles FEC, Modulation, mapping to the physical channels, etc.

a) RRC Sublayer The main functions of the RRC sublayer include

- Broadcast of System Information related to the non-access stratum (NAS) and access stratum (AS)
- Paging
- Establishment, maintenance, and release of an RRC connection between the UE and E-UTRAN (allocation of temporary identifiers)
- Security functions including key management
- Establishment, configuration, maintenance, and release of Radio Bearers

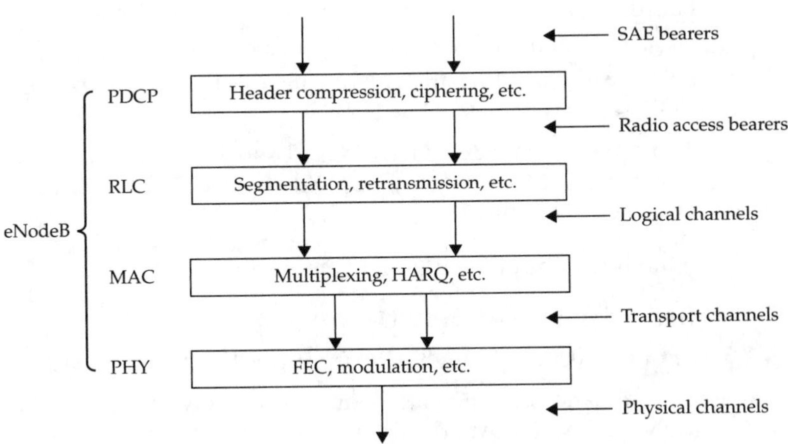

FIGURE 2-34 LTE user-plane protocol architecture

- Mobility functions, including UE measurement reporting, handover UE cell selection and reselection, etc.
- Establishment, configuration, maintenance, and release of Radio Bearers for MBMS services
- QoS management functions

b) PDCP Sublayer The main functions of the PDCP sublayer for the user plane include

- Header compression and decompression
- Transfer of user data
- In-sequence delivery of upper-layer PDUs at PDCP re-establishment procedure for RLC AM
- Retransmission of PDCP SDUs at handover for RLC AM
- Ciphering and deciphering
- Timer-based SDU discard in uplink

The main functions of the PDCP for the control plane include

- Ciphering and integrity protection
- Transfer of control plane data

c) RLC Sublayer The RLC handles retransmissions of data, segmentation and concatenation of data, and in-sequence delivery of data to higher layers. As with WCDMA, there are three modes of configuration:

- **Acknowledged Mode (AM)** In this case the RLC performs retransmission of erroneous data. This is typically used in packet data services requiring error-free delivery.
- **Un-acknowledged Mode (UM)** Handles in-sequence delivery but no retransmission of data; this is typically used in VOIP services requiring low latency.
- **Transparent Mode (TM)** Used in random access. Transparent Mode is only applied to BCCH, CCCH, and PCCH.

The main functions of the RLC sublayer include

- Transfer of upper-layer PDUs
- Error Correction through ARQ (only for AM data transfer)
- Concatenation, segmentation, and reassembly of RLC SDUs (only for UM and AM data transfer)
- Resegmentation of RLC data PDUs (only for AM data transfer)

- In-sequence delivery of upper-layer PDUs (only for UM and AM data transfer)
- Protocol error detection and recovery

d) MAC Sublayer The MAC handles HARQ retransmissions, multiplexing, scheduling, etc. The serving cell is defined as the cell the mobile terminal is communicating with and is responsible for HARQ functions. The MAC prepares data to be transmitted and places them into transport blocks (TB). A transport format conveys information on how the TB is transmitted (such as TB size, modulation scheme, etc.). As discussed earlier, the scheduling function handles the assignment of resources; it is located in the MAC.

The main functions of the MAC sublayer include

- Mapping between logical channels and transport channels
- Multiplexing/demultiplexing of MAC SDUs belonging to one or different logical channels into/from transport blocks (TB) delivered to/from the physical layer on transport channels
- Scheduling information reporting
- Error correction through HARQ
- Priority handling between logical channels of one UE
- Priority handling between UEs by means of dynamic scheduling
- Transport format selection

e) PHY Sublayer This physical sublayer converts the transport channels into physical channels. On the downlink one or two transport blocks can be transmitted per TTI; the scheduler determines the modulation choice, FEC rate, etc. On the uplink one transport block can be transmitted per TTI.

In referring to the protocol architecture provided in earlier Figure 2-34, we will now describe the logical, transport, and physical channel types for both uplink and downlink.

The Logical Channel Types are

- **Paging Control Channel (PCCH)** Used to transmit paging information to UEs
- **Broadcast Control Channel (BCCH)** Used to transmit system control information to UEs
- **Multicast Control Channel (MCCH)** Used to transmit point-to-multipoint MBMS control information to receive MTCH
- **Multicast Traffic Channel (MTCH)** Used for point-to-multipoint MBMS traffic data

- **Dedicated Control Channel (DCCH)** Used to transmit point-to-point bidirectional control information
- **Dedicated Traffic Channel (DTCH)** Used to transmit dedicated point-to-point user data

The Transport Channel Types are

- The downlink transport channel types are
  - **Broadcast Channel (BCH)** Broadcasting a predefined transport format in the entire coverage area of the cell
  - **Downlink Shared Channel (DL-SCH)** Support for HARQ, dynamic link adaptation, and MBMS transmission
  - **Paging Channel (PCH)** Broadcast in the entire coverage area of the cell
  - **Multicast Channel (MCH)** Support for MBSFN combining of MBMS transmission on multiple cells
- Uplink transport channel types are
  - **Uplink Shared Channel (UL-SCH)** Support for dynamic link adaptation and HARQ
  - **Random Access Channel (RACH)** Limited control information with collision risk

The Physical Channel Types are

- **Physical broadcast channel (PBCH)** The coded BCH transport block is mapped to four subframes within a 40 ms interval.
- **Physical control format indicator channel (PCFICH)** Informs the UE about the number of OFDM symbols used for the PDCCHs.
- **Physical downlink control channel (PDCCH)** Informs the UE about the resource allocation of PCH and DL-SCH, and Hybrid ARQ information related to DL-SCH.
- **Physical Hybrid ARQ Indicator Channel (PHICH)** Carries Hybrid ARQ ACK/NAKs in response to uplink transmissions.
- **Physical downlink shared channel (PDSCH)** Carries the DL-SCH and PCH.
- **Physical multicast channel (PMCH)** Carries the MCH.
- **Physical uplink control channel (PUCCH)** Carries HARQ ACK/NAKs in response to downlink transmission; carries Scheduling Request (SR) and CQI reports.
- **Physical uplink shared channel (PUSCH)** Carries the UL-SCH.

- **Physical random access channel (PRACH)** Carries the random access preamble.

The scheduler will use both frequency and time domains in assigning resources, the scheduler utilizes information about the downlink by monitoring the uplink CQI reports from the UE. Here the UE measures downlink reference channels.

The HARQ is located in the MAC sublayer. Certain channel types are eligible for HARQ transmission; they are UL-SCH and DL-SCH. Since there is a possibility the transport block will be correctly decoded out of sequence, the received transport blocks are de-multiplexed into logical channels and the RLC will re-order them based on the sequence number assigned.

One point worth mentioning is for LTE the HARQ re-transmissions and RLC re-transmissions are both located in the eNodeB. While for HSPA, the RLC re-transmission is located in the RNC and the HARQ is located in the NodeB. Since these functions are co-located in LTE, a close relationship can build with the goal of improving the overall system performance (i.e., reducing system latency, etc.).

LTE Network Functionality Partitioning

Next we provide a brief overview of the functions performed in the eNodeB and the evolved packet core (EPC). This presentation provides a slightly different perspective in order to fully capture the LTE benefits. Earlier we have provided functional definitions based on the protocol layers; now we provide the functional partitioning based on the actual network architecture location.

The *eNodeB* hosts the following functions:

- Functions for Radio Resource Management: Radio Bearer Control, Radio Admission Control, Dynamic UL/DL allocation of resources to UEs (scheduling)
- IP header compression and encryption of user data stream
- Selection of an MME at UE attachment
- Routing of user-plane data toward Serving Gateway
- Scheduling and transmission of paging messages and broadcast
- Measurement and measurement reporting configuration for mobility and scheduling

The *Mobility Management Entity (MME)* hosts the following functions:

- NAS signaling and signaling security
- AS Security control

- Inter–CN node signaling for mobility between 3GPP access networks
- PDN GW and Serving GW selection
- SGSN selection for handovers to 2G or 3G 3GPP access networks
- Roaming
- Authentication
- Bearer management functions, including dedicated bearer establishment

The *Serving Gateway (S-GW)* hosts the following functions:

- The local Mobility Anchor point for inter-eNodeB handover
- Mobility anchoring for inter-3GPP mobility
- E-UTRAN idle mode downlink packet buffering and initiation of network triggered service request procedure
- Lawful interception
- Packet routing and forwarding
- Transport-level packet marking in the uplink and the downlink
- Accounting on UL and DL charging

The *PDN Gateway (P-GW)* hosts the following functions:

- Per-user based packet filtering (by, e.g., deep packet inspection)
- Lawful interception
- UE IP address allocation
- Transport-level packet marking in the downlink
- UL and DL service-level charging, gating, and rate enforcement

This functionality partitioning is drawn in Figure 2-35, where we have specifically called out the S1 user and control plane interfaces.

For LTE the downlink scheduler performs similar functions as the HSDPA, except now time and frequency domains are used in the scheduling of users. The uplink is where LTE and HSUPA scheduling differ: in HSUPA the NodeB sets an upper limit on the amount of interference the UE can impose into the system and lets the UE select its transport format; the UE is therefore required to signal this information to the NodeB. However, for LTE, the story is different, since orthogonality can be preserved on the uplink. In this case the eNodeB is responsible for the transport format of the UE and thus no signaling is required to the eNodeB. This is a prudent choice of techniques, since the uplink of HSPA and LTE have different system-related properties.

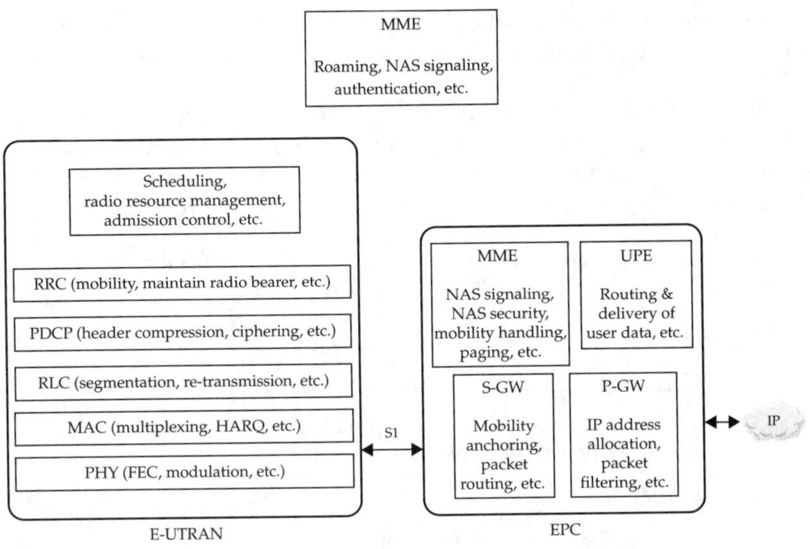

FIGURE 2-35 Functionality partitioning between E-UTRAN and EPC

Earlier in this chapter we presented the WCDMA and HSDPA network architecture, it consisted of the RAN (NodeB and RNC) and CN. The LTE network architecture given in Figure 2-36 consists of the eNodeB and EPC. In fact, this topology shows there is no uplink or downlink macro-diversity support. This is not to say that handoffs don't exist, as they should to accommodate the mobility requirements for LTE.

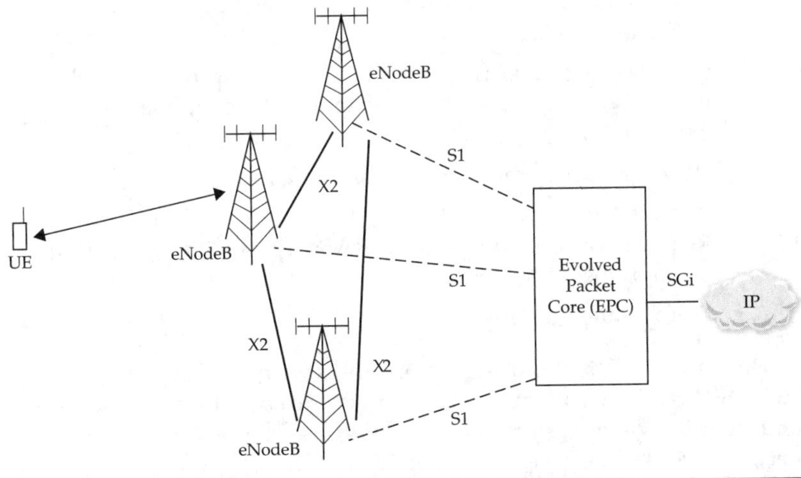

FIGURE 2-36 LTE network architecture

Here the eNodeB has taken on most of the RNC functionality where the S1 interface is similar to Iu and the X2 interface is similar to Iur.

The eNodeB is essentially equivalent to the NodeB and RNC. The eNodeB handles handoff uplink and downlink scheduling, radio resource management, etc. There is no concept of a serving or drift eNodeB in LTE. Handoffs are accomplished by eNodeB relocations.

X2 Interface The list of functions on the X2 interface is the following:

- Intra LTE-Access-System Mobility Support
  - Context transfer from source eNodeB to target eNodeB
  - Control of user plane transport bearers between source eNodeB and target eNodeB
- Load management
- Intercell interference coordination
  - Uplink interference load management
- General X2 management and error handling functions
- Application-level data exchange between eNodeBs

S1 Interface The radio network signaling over S1 consists of the S1 Application Part (S1AP). The S1AP protocol consists of mechanisms to handle all procedures between the EPC and E-UTRAN. It is also capable of conveying messages transparently between the EPC and the UE without interpretation or processing by the E-UTRAN.

Over the S1 interface the S1AP protocol is, e.g., used to

- Facilitate a set of general E-UTRAN procedures from the EPC such as paging-notification as defined by the notification SAP.
- Separate each User Equipment (UE) on the protocol level for mobile-specific signaling management as defined by the dedicated SAP.
- Transfer transparent non-access signaling as defined in the dedicated SAP.
- Request various types of E-RABs (provided by the AS) through the dedicated SAP.
- Perform the mobility function.

In Figure 2-37, for comparison purposes we provide the WCDMA and HSPA core network in a simplified view to show that both packet- and circuit-switched services exist. We also show how the GSM network would interface to this network as well.

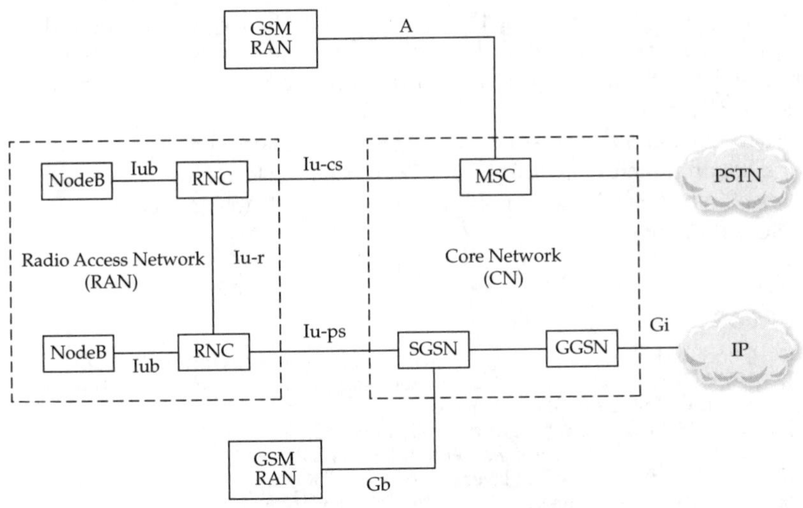

FIGURE 2-37 Simplified WCDMA and HSPA Core Network view

Here, the following definitions apply: Serving GPRS Support Node (SGSN), Gateway GPRS Support Node (GGSN), and Mobile Switching Center (MSC).

The evolved packet core is also referred as the SAE and only the packet data services are addressed here. Recall the NGN will support an all-IP-based network topology. In Figure 2-38 we provide

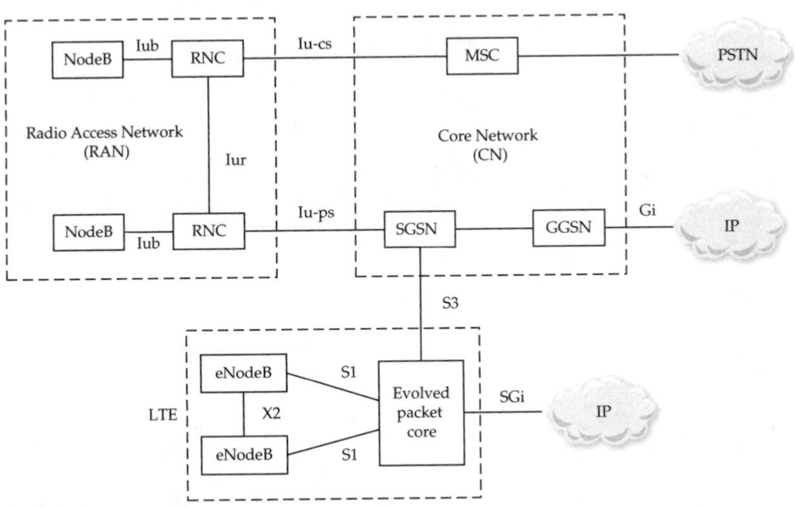

FIGURE 2-38 SAE Core Network view

an example of interfacing the LTE network to the WCDMA and HSPA network. The initial deployments of the LTE RAT will need to coexist with the 3G networks until all the traffic can be placed on the IP domain.

The SAE only address the packet switch domain, since the concept of a flat architecture is utilized in LTE. In LTE the RAN is the eNodeB; hence, in order to handle mobility, the SAE needs to connect to every eNodeB in the network.

References

[1] J. Boccuzzi, *Signal Processing for Wireless Applications*, Mc-Graw Hill, 2007.
[2] 3GPP TS 25.211: "Physical channels and mapping of transport channels onto physical channels (FDD)."
[3] 3GPP TS 25.212: "Multiplexing and channel coding (FDD)."
[4] 3GPP TS 25.213: "Spreading and modulation (FDD)."
[5] 3GPP TS 25.214: "Physical layer procedures (FDD)."
[6] 3GPP TS 25.215: "Physical layer – Measurements (FDD)."
[7] 3GPP TS 36.201: "Evolved Universal Terrestrial Radio Access (E-UTRA); Physical Layer – General Description."
[8] 3GPP TS 36.212: "Evolved Universal Terrestrial Radio Access (E-UTRA); Multiplexing and channel coding."
[9] 3GPP TS 36.213: "Evolved Universal Terrestrial Radio Access (E-UTRA); Physical layer procedures."
[10] 3GPP TS 36.214: "Evolved Universal Terrestrial Radio Access (E-UTRA); Physical layer – Measurements."
[11] 3GPP TS 36.104: "Evolved Universal Terrestrial Radio Access (E-UTRA); Base Station (BS) radio transmission and reception."
[12] 3GPP TS 36.101: "Evolved Universal Terrestrial Radio Access (E-UTRA); User Equipment (UE) radio transmission and reception."

CHAPTER 3

Femtocell System Analysis

In this chapter we will present a foundation to support system analysis of femtocell deployments. We begin with a discussion on the various deployment scenarios expected to be used. This will also cover an introduction to the handoffs expected to occur between the public macrocell and the private femtocell. Next a discussion of the indoor path loss models is provided. These path loss models compare the various options available to the system designer. Immediately following this discussion, we offer examples of uplink and downlink link budgets. A simple capacity example is provided. The 3GPP RF requirements corresponding to the Home NodeB is reviewed, and finally implementation-related issues are addressed.

3.1 Deployment Scenarios

Figure 3-1 presents femtocell reference architecture diagram. The femtocell is communicating wirelessly to the UE and is connected to the modem to provide broadband Internet access.

It should come as no surprise that certain deployment scenarios and configurations will create interference to either the uplink or the downlink. For example, access to the femtocell is critical and requires careful definition. You will have either open-access or closed-access configurations. For the open-access case, the femtocell is allowed to provide service to any UE within its coverage area. This is pretty much the same approach that the macrocell NodeB supplies. Visitors, nearby pedestrians, etc., have access to the femtocell and are allowed to access the cellular service. There exists an upper limit to the number of simultaneous users that can be serviced, as limited by the femtocell equipment, broadband connection bandwidth, amount of interference, etc.

For the closed-access case, the femtocell is only allowed to provide service to a particular group of UEs, called the Closed Subscriber Group (CSG). This small group of UEs will be determined through some negotiation between the femtocell owner and the cellular

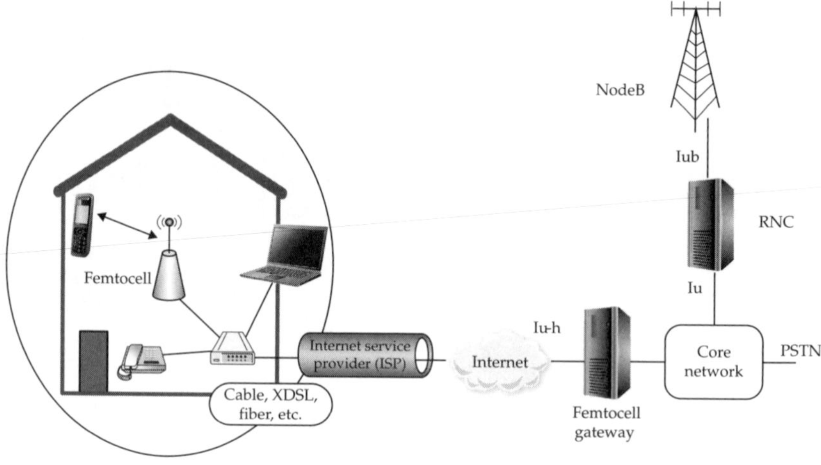

FIGURE 3-1 Femtocell reference architecture block diagram

service provider. We also envision the femtocell owner to have the capability to allow certain users to connect to the femtocell. This can be accomplished through a robust user interface, either through an application running on a laptop or desktop computer or the primary cell phone. Similar limits exist as to the number of UEs that can access the femtocell service.

As previously discussed, it is not feasible to expect deployments without interference. This is a concern not only to the femtocell, but also to the NodeB. Please recall the overall goal is to provide improved coverage in the home. Since complete control over the placement of the femtocells may not be possible, various interference scenarios have been studied in the 3GPP standards group. These scenarios are listed in Table 3-1.

Figure 3-2 serves as a companion figure to Table 3-1, which is used to better explain the source of interference. The first scenario is

Number	Aggressor	Victim
1	UE attached to Home NodeB	Macro NodeB Uplink
2	Home NodeB	Macro NodeB Downlink
3	UE attached to Macro NodeB	Home NodeB Uplink
4	Macro NodeB	Home NodeB Downlink
5	UE attached to Home NodeB	Home NodeB Uplink
6	Home NodeB	Home NodeB Downlink

TABLE 3-1 Femtocell Interference Scenarios

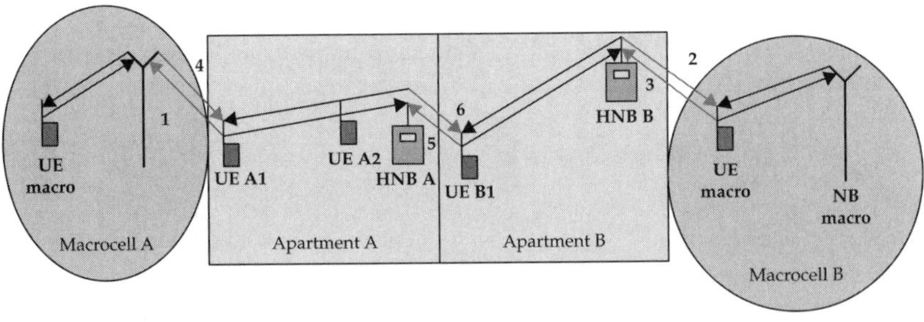

FIGURE 3-2 Interference scenario diagram

when a UE connected to a Home NodeB causes interference to the uplink of a public NodeB. This interference can be found when both the public and private cells are using the same frequency as well as when they are adjacent. The second scenario corresponds to the Home NodeB causing interference to the downlink of the public cell. The UE connected to the public NodeB will fall victim to this. The third scenario is when the UE connected to the public NodeB is close to the Home NodeB. Here the uplink of the Home NodeB will observe a rise in interference, since the UE needs to transmit with higher power to reach the public NodeB. The fourth scenario occurs when the downlink of the public cell interferes with the downlink of the Home NodeB. In the fifth scenario a UE connected to one Home NodeB interferes with the uplink of another, nearby Home NodeB. Finally, the sixth scenario is when the downlink of one Home NodeB interferes with the downlink of another, nearby Home NodeB.

Please note coexistence with other technologies such as WiMax or CDMA2000 is not covered in this description but should be considered, depending on the frequency allocations.

3.1.1 Handoff Discussion

Since hard handoffs are supported within the femtocell standard, we now address the possible handoff scenarios of interest. Specifically we mention handoffs from the public cell to the private cell, as shown in Figure 3-3. This use case would resemble the situation when one is returning to the home environment. Here the UE will make measurements of neighboring cells, both private and public, and then report them (via periodic or aperiodic intervals) to the NodeB and RNC. The RNC will recognize the femtocell ID and route the phone call conversation to the home environment. In this case the femtocell is located within the user's home and both the UE and RNC are aware of this configuration. In order to provide a reliable mechanism, it is envisioned that a geographically specific table of femtocell IDs (among other identifiers) would be stored to assist in these types of decisions.

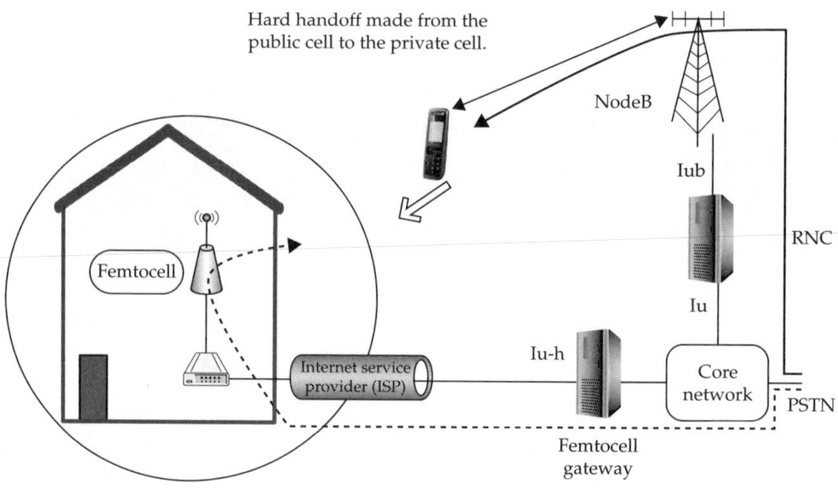

Figure 3-3 Hard handoff from the public cell to the private cell

Also, handoffs from the private cell to the public cell are possible, as is highlighted in Figure 3-4. This use case would resemble the situation when the phone call was originated within the home environment and then the user is leaving its premises. Here the UE will make measurements of neighboring cells, both private and public and then report them (via periodic or aperiodic intervals) to the femtocell. The femtocell will alert the cellular network to route the phone call conversation to the public cell. In order to provide a reliable mechanism,

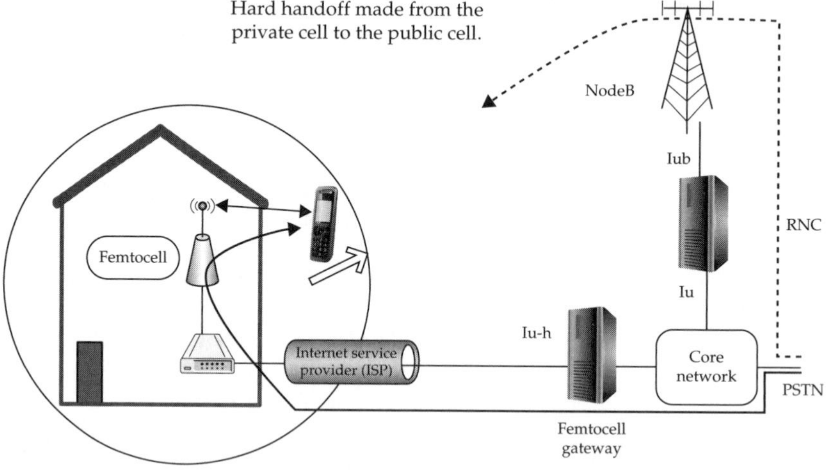

Figure 3-4 Hard handoff from the private cell to the public cell

it is envisioned that a geographically specific table of femtocell IDs (among other identifiers) would be stored to assist in these types of decisions.

3.2 Path Loss Models

This section will provide an overview of the path loss models available to the system designer. Unlike the outdoor propagation phenomenon, indoor propagation is more involved due to the presence of the multiple walls and floors encountered in the propagation path.

3.2.1 Path Loss Model #1

The specific indoor model comes to us via the ITU-R standard, specifically the ITU-R P.1238. The path loss model is given by the following equation, whose value is expressed in units of dB:

$$L_1 = 20\log(f_c) + 10n\log(r) + L_F(n_F) - 28 \tag{3.1}$$

Here, f_c is the carrier frequency expressed in MHz, n is the path loss exponent, r is the distance separation expressed in meters, and L_F is the floor penetration loss, which varies by the number of penetrated floors, n_F.

Let's consider the following scenario of interest: no floors are being considered and the carrier frequency is $f_c = 2.4$ GHz. The resulting path loss model resembles the following:

$$L_1 = 39.6 + 10n\log(r) \tag{3.2}$$

A wide variety of public literature shows the path loss exponent can vary wildly, depending on the environment (residential, commercial, etc.), carrier frequency, etc.

We provide some insight later in this subsection; however, suffice it to say the path loss exponent can vary from 2.0 to approximately 3.8.

The preceding equation can be extended to consider the impact of multiple floors as shown here:

$$L_{1B} = 39.6 + 10n\log(r) + 15 + 4(n_F - 1) \tag{3.3}$$

Figure 3-5 compares the path loss models at 2.4 GHz. The plot labeled L1 shows the path loss versus distance for this model when the transmit and receive devices are located on the same floor. The additional plots labeled L1B and L1C correspond to having one and two floors between the transmit and receive devices. As one can see, once the first floor is included in the calculation, the loss has increased significantly. From Equation 3.3, we can conclude the addition of a single floor increases the path loss by 15 dB, while including two floors increases the loss by 19 dB.

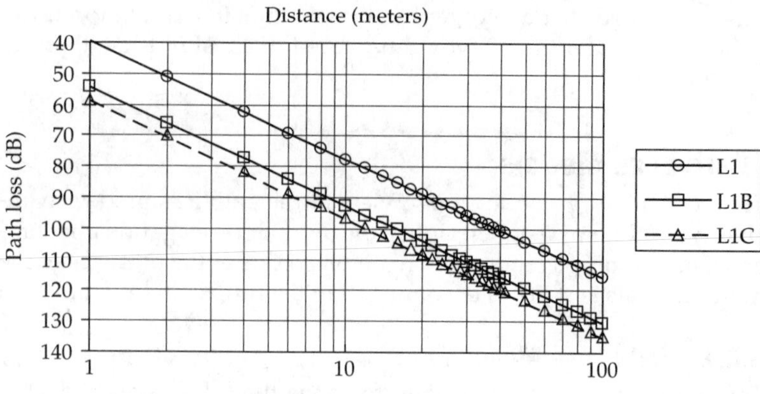

Figure 3-5 Path loss model 1 propagation plot

3.2.2 Path Loss Model #2

Another useful path loss model from [1] is given next for 900 MHz carrier frequency. Table 3-2 shows both upper and lower limits to consider the randomness of the wireless environment. The variable r is used to denote the spatial separation in distance.

A plot of the path loss model is provided in Figure 3-6. One can see how the slope of the propagation path loss phenomenon increases in segments, thus significantly further decreasing the received signal power.

3.2.3 Path Loss Model #3

This next case is the Cost 231 model, where further information can be found in [2]. This model defined various wall types, hence thin and thick walls would be able to be distinguished. Moreover, the loss due to additional floors is different from the previously presented model.

$$L = 37 + 20\log(r) + \sum_{j=1}^{N_w} L_{wj}\alpha_{wj} + L_F N_F^{\left[\frac{N_F+2}{N_F+1}-0.46\right]} \tag{3.4}$$

Distance (Meters)	Lower Path Loss	Upper Path Loss
1 < r < 10	30 + 20log(r)	30 + 40log(r)
10 <= r < 20	20 + 30log(r)	40 + 30log(r)
20 <= r < 40	–19 + 60log(r)	1 + 60log(r)
40 <= r	–115 + 120log(r)	–95 + 120log(r)

Table 3-2 Ericsson Indoor Propagation Path Loss Model

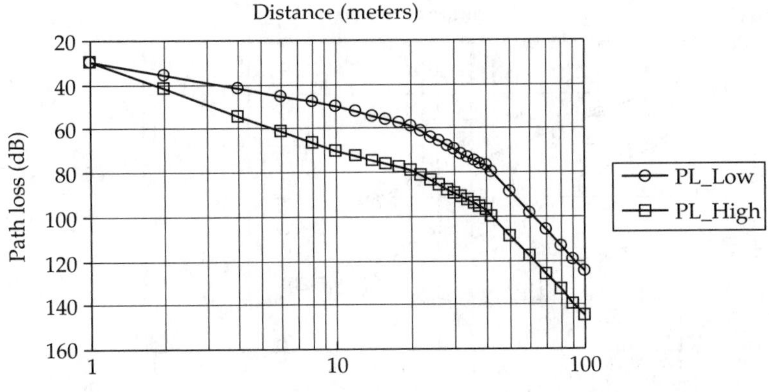

FIGURE 3-6 Path loss model #2 propagation plot

Here N_w is the number of wall types, L_{wj} is the wall loss for type j, α_{wj} is the number of walls encountered for type j, L_F is the floor loss, and N_F is the number of floors encountered. Various information is available for the preceding parameters; in Table 3-3 we list two of them.

Please notice that the floors have higher attenuation than the walls. If we consider a single inside wall and a single floor to overcome, then the following path loss model is applicable for 1800 MHz:

$$L_{3B} = 37 + 20\log(r) + 3.4 + 18.3 \tag{3.5}$$

And the following would apply for 900 MHz:

$$L_{3C} = 31 + 20\log(r) + 1.9 + 14.8 \tag{3.6}$$

The plots in Figure 3-7 show the predicted path loss when considering a single floor and additional wall to be encountered in the transmission path. The two curves correspond to 900 MHz and 1800 MHz carrier frequencies. The higher carrier frequency corresponds to the line with the larger propagation path loss. Moreover, we can easily see as the wireless signal encounters not only the outer wall, but also a few floors of obstruction, the aggregate path loss can be significant.

TABLE 3-3 Wall and Floor Path Loss Examples

Parameter	900 MHz	1800 MHz
L_W	1.9 dB	3.4 dB
L_F	14.8 dB	18.3 dB

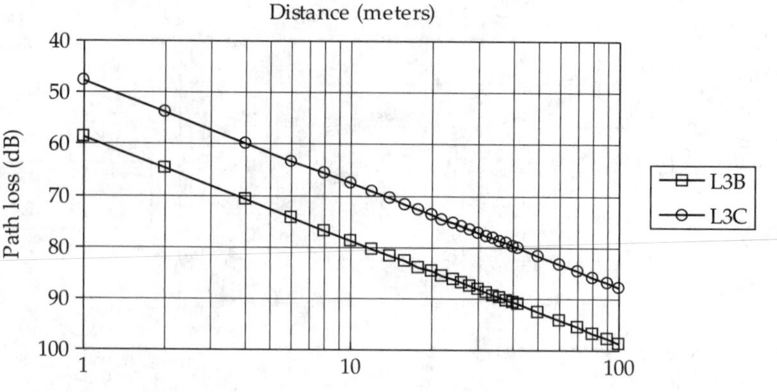

FIGURE 3-7 Path loss model #3 propagation plot

3.2.4 Path Loss Model #4

This next model is based on measurements conducted at 5 GHz by [3], who have separated measurements into various categories such as residential and commercial, as well as whether obstructions were within the propagation path. We have focused on the results for the residential premises. The line of sight (LOS) measurements are provided first:

$$L_{4A} = 45.9 + 20.1\log(r) + 3.2 \tag{3.7}$$

They are followed by the non–line of sight (NLOS) results:

$$L_{4B} = 50.3 + 31.2\log(r) + 3.8 \tag{3.8}$$

A plot comparing the path loss difference between the two models is shown in Figure 3-8. This difference translates into 15 dB at a distance of 10 meters of separation and increases as the distance increases.

3.2.5 Path Loss Model #5

In this example measurements were made in a laboratory environment at 1800 MHz [4]. What is particularly interesting with these measurements is they have shown the path loss exponent to vary as the number of floors traversed is increased (see Table 3-4). The previous models had simply added a constant to keep the exponent the same. The results show the impact of floor path loss on the exponent.

$$L_5 = 34.7 + 10n\log(r) + \sigma \tag{3.9}$$

Table 3-4 Path Loss Exponent Parameters

Scenario	Exponent	Shadowing
Same floor	4.5	8.7
Through 1st floor	5.3	2.9
Through 2nd floor	5.6	3.0
Through 3rd floor	6.3	4.7

A plot comparing all of these path loss models is presented in Figure 3-9. When considering a distance of 10 meters, the additional loss is 2.2 dB, 5.3 dB, and 14 dB for one, two, and three floors, respectively.

3.2.6 Path Loss Comparison

In this subsection we provide a comparison of some of the previously mentioned path loss models. Specifically, some of the salient features are emphasized. The first comparison shows the impact of adding floor loss to the overall path loss model, assuming 900 MHz carrier frequency (see Figure 3-10). Here we see the two models compared can differ by greater than 5 dB. Also, we notice how as more floors are considered, the incremental loss decreases. These models added a constant (or appropriately shifted the path loss plot), depending on the number of floors considered.

Figure 3-11 compares some of these path loss models, assuming 900 MHz carrier frequency. For these three particular models, we can see how the models correlate very well at up to 20 meters but deviate past this distance. This is because of the inclusion of floor/wall losses in one of the path loss models, specifically L2. In any event the

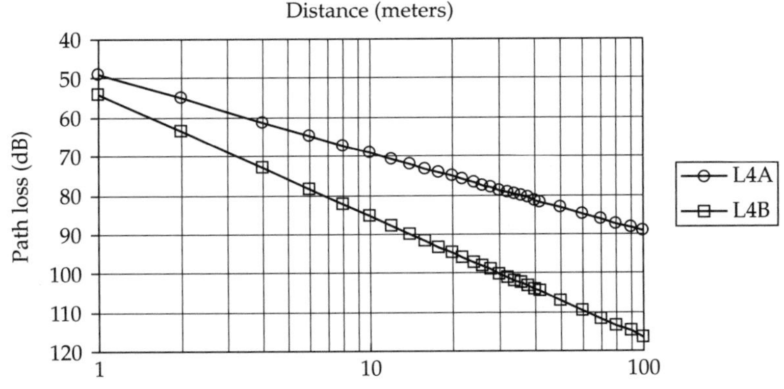

Figure 3-8 Path loss model #4 propagation plot

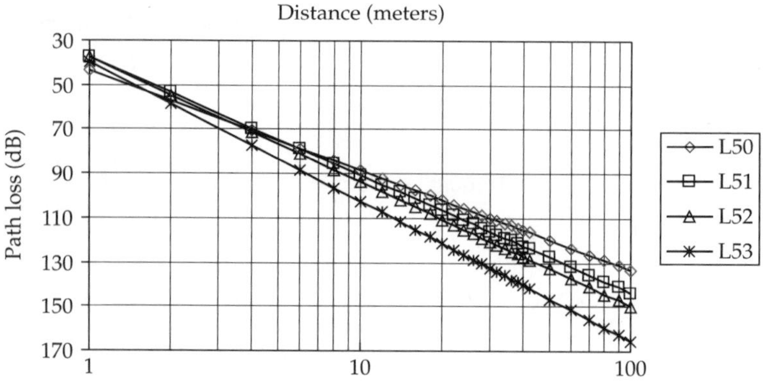

FIGURE 3-9 Path loss model #5 propagation plot

potential path loss can be seen to be significant—and should be accurately considered in the link budget calculation.

A point worth making here is that we have just presented two methods to include the impact of floor loss. One method simply adds a constant loss, keeping the slope the same, while the other method adjusts the slope while keeping the constant reasonably the same. Both approaches will increase the path loss, while only the latter will significantly increase the path loss as the distance is increased. Hence caution should be used when selecting a model; it is always best to collect some measurements within the targeted deployment sites, simply because the actual construction of the buildings can vary significantly and hence so can the path loss. The reader is encouraged to visit the following references for additional technical material: [5], [6], [7], and [8].

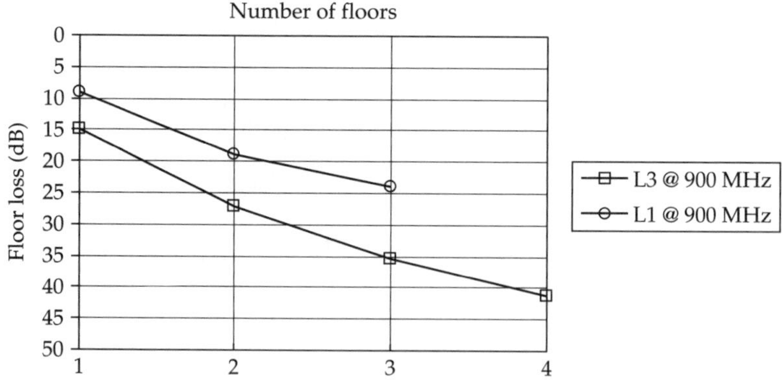

FIGURE 3-10 Floor loss comparison of some path loss models

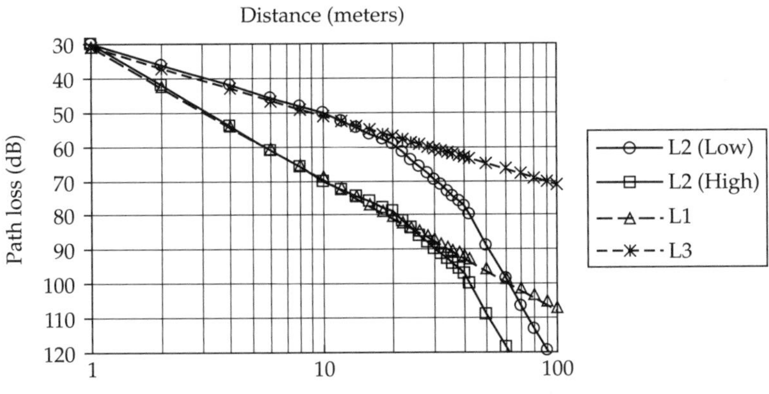

FIGURE 3-11 Path loss comparison of some models

3.3 Link Budget

Next link budget is provided for the femtocell scenario, where we have considered the case when the UE is within the home. Table 3-5 provides the parameter listing and values used in the analysis, along with their units.

In the table we have segmented certain sections of the link budget. The Transmit Power parameter (label a) corresponds to the femtocell transmit power. We have assigned this to 0 dBm (or 1 milliWatt) for this example. For the sake of completeness, we have also included Transmit Cable Loss (label b) and Antenna Gain (label c). For femtocell packages with the antenna built into the product, this would not be relevant, and hence these parameters should be set to a value of 0. This is the situation we decided to follow, but we left a placeholder here for cases when an external antenna would be used for sake of deployment conditions (such as a small campus) and/or limitations. We have also forced the antenna gain field to zero dB, assuming a more pessimistic approach; however, such a pessimistic approach may be warranted depending on the particular implementation. The same philosophy was applied on the receiver with the Cable Loss (label e) and Antenna Gains (label d). We have assumed the UE communicating to the femtocell; however, one should not preclude other, possibly less mobile devices from being connected to the femtocell.

Thermal noise (label f) and interference noise (label g) densities were addressed next. We have assumed the interference from the public macrocell and neighboring femtocells operating on the same frequency band produces approximately 5 dB noise rise (label t) in the system. There is much flexibility in the table to increase this value to also consider the other UEs communicating simultaneously within the femtocell. However, we felt that 5 dB is a reasonable assumption.

Label	Parameter	Value	Unit	General Comments
a	Transmit Power	0	dBm	1 mWatts
b	Transmit Cable Loss	0	dB	
c	Transmit Antenna Gain	0	dBi	
d	Receiver Antenna Gain	0	dBi	
e	Receiver Cable Loss	0	dB	
f	Thermal Noise Density	–174	dBm/Hz	3.98E-18
g	Interference Density	–170.5	dBm/Hz	8.91E-18
h	Total Inter + Noise Density	–168.90	dBm/Hz	–168.90
i	Receiver Noise Figure	9	dB	
j	System Band Width	65.84	dB	
k	Processing Gain	21.07		
l	Lognormal Shadowing	6	dB	
m	Propagation Path Loss	104	dB	May include floor loss
n	Floor/Wall Path Loss	0	dB	
o	Various Implementation Losses	0	dB	
p	**Rx Signal Power**	**–110**	dBm	a – b + c + d – e – l – m – n – o
q	**Eb/(Io+No)**	**6.90**	dB	p – h – i – j + k + 1.77
r	**Ec/(Io+No)**	**–15.95**	dB	p – s
s	**Interference + Noise Floor**	**–94.05**	dBm	h + i + j
t	**Noise Rise Over Thermal**	**5.10**	dB	h – f

Table 3-5 Femtocell Link Budget Example

The UE receiver noise figure (label i) is assumed to be 9 dB; this is obviously implementation dependent due to design trade-offs between performance, cost, etc. This should be a reasonable upper target for the UE; in fact smaller values should be used.

The system considered here is the 3G WCDMA Radio Access Technology (RAT), which has a chip rate of 3.84 MCps and is coined System Bandwidth (label j). The processing gain (label k) of the WCDMA system is determined by the number of chips used to spread the modulation symbol.

Next the channel propagation loss section is encountered in the link budget. The Shadowing value (label l) has been set to 6 dB in this example. The Propagation Path Loss (label m) and Floor/Wall Loss (label n) fields are set to 104 dB and 0 dB, respectively. The reason for the specific path loss value will be discussed shortly, but let's discuss why the Floor and Wall Loss have been separately called out in the link budget. The Floor/Wall Loss is a separate row because some models add a constant value to consider the effects of the floor or wall penetration. In this case, it would be reasonable to expect to place that value in this location. Recall other models simply include the additional penetration loss by modifying the exponent; in this case, however, the Floor/Wall loss cell should be set to 0 dB in order not to count that loss into the system calculation twice.

A short discussion on the Various Implementation Losses (label o) field is appropriate. The analysis so far would predict the received signal power, which would then be used to estimate SNR of some sort. Since other impairments such as ADC quantization noise, residual frequency offset, and residual timing offset degradations are not directly considered, we have left a placeholder for them if the designer so chooses to include them into this link budget for a more realistic view. One last comment to this point, we expect the receivers to utilize advanced signal processing techniques to not only mitigate multipath, but also interference.

Next we enter the derived section of the link budget. We begin with the Received Signal Power (label p). This is essentially all the system losses subtracted from the system gains. In other words, the Transmit Antenna Gain is added to the Transmit Power, but the cable loss is subtracted from the resultant. We have placed the exact fields used in deriving the estimate in the general comments column. For this case chosen, the received signal power is –110 dBm. The interference plus noise floor is calculated to be a value of –94.05 dBm.

As previously discussed, we have strategically chosen the path loss and shadowing combination in order to achieve the $E_c/(I_o + N_o)$, which is label r, with a value of approximately –16 dB. This value defines the chip energy ratio to the sum of the interference and noise density. Some designers prefer to operate with bit energies; hence, we have specifically created a row called $E_b/(I_o + N_o)$, which is label q.

This value of approximately 6.9 dB corresponds to the bit energy ratio to the interference and noise density. This bit that we are referring to corresponds to bits after demodulation and after Forward Error Correction (FEC) decoding. A point worth mentioning is we targeted an $E_c/(I_o + N_o)$ target of –16 dB, since this is sufficient for voice calls. As indicated in Table 3-5, this was obtained assuming a total propagation loss of 104 + 6 = 110 dB. Recall the path loss models presented earlier, this range of values corresponds to distances in the range of greater than 50 meters (depending on the carrier frequency chosen) and the number of walls and floor considered. Finally, we have assumed a 1 mWatt transmitter.

As previously mentioned, supplying a high SNR indoors from a public NodeB that is physically located a few kilometers away is a challenging task. One of the reasons is overcoming the outer wall penetration loss. Next we attempt to provide a simple viewpoint of where the outer wall will work to the advantage of the femtocell when addressing the leakage of interference from neighboring femtocells. Figure 3-12 shows a public NodeB path loss, shown as a solid line, to the home. Once this signal enters the home, a loss of anywhere from 10 dB to 20 dB can be found (depending on the building material) and thus degrade and also prevent service. The femtocell propagation path, shown as a dashed line, will also encounter the outer wall penetration loss. However, the signal power or SNR available within the home will be larger than that from the public macrocell, which is located a few kilometers away.

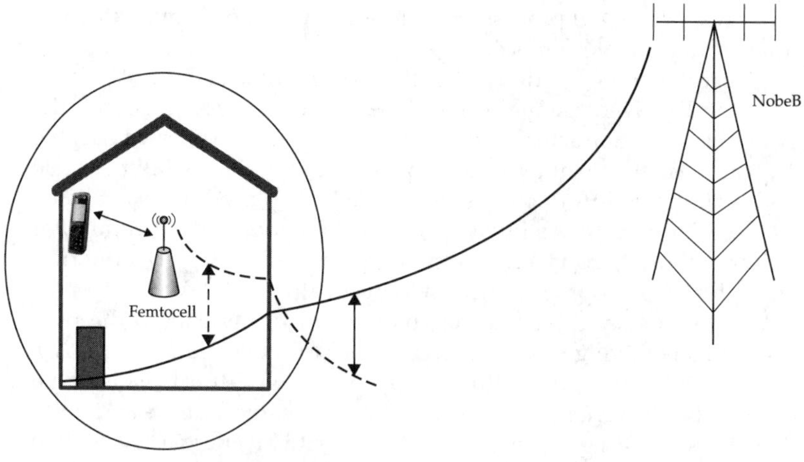

Figure 3-12 Femtocell and macrocell propagation summary

3.4 Example Capacity

Considering certain power allocations for the HSDPA services, we have the following example use case. Capacity in general is a very involved topic, especially with the assumptions used in determining the number of users the cell is capable of servicing. The approach we have taken is simple, yet robust. We have assigned various power levels to the physical channels transmitted by the femtocell. The remaining power allocation is assumed to be consumed by additional users. This will provide the maximum use of the femtocell transmit power.

Table 3-6 describes the WCDMA user capacity example of interest. We have chosen certain power allocations (E_c/I_{or}) for the various downlink physical channels. We have also placed the percentage of the total transmitted power next the values. The total sum should equal one. The sum of the physical channels is subtracted from 1 and the remaining power is typically allocated to OCNS in simulations, but here we have assumed other WCDMA voice users would easily consume the remaining power. With the following power allocations, approximately seven users would be able to operate in the femtocell. However, to allow for some head room for power control, etc., we can say approximately four users can be serviced. Please note that HSDPA and HSUPA are assumed to be operational, taking up as much as 60 percent of the total transmitted power.

Channel	Ec/Ior dB	Percentage	Ec/Ior dB	Percentage
P-CPICH	−10	10	−10	10
P-CCPCH	−12	6.31	−12	6.3
SCH	−12	0	−12	0
PICH	−18	1.58	−18	1.58
DPCH	−16	2.51	−16	2.51
HS-DSCH	−3	50.12	−2	63.1
HS-SCCH	−10	10	−9	12.6
E-HICH	−22	0.63	−22	0.63
E-RGCH	−20	1	−20	1
E-AGCH	−20	1	−20	1
OCNS	−7.74	16.84	−18.9	1.28
WCDMA Capacity		7.7		1

TABLE 3-6 Example of Femtocell Power Allocations

The power allocations in the next column correspond to an all-HSDPA system, meaning the UE would communicate via HSDPA and HSUPA channels not only for data, but also for voice applications such as VoIP. Here HSDPA is taking up as much as 75 percent of the total transmitted power. In this case there will be a single, high data rate, shared resource among the femtocell users. Hence, depending on what the use case distribution is—for example, all voice call users, some data users, or high speed users—the number of serviceable users varies.

3.5 3GPP RF Specifications

The addition of the femtocell or Home NodeB to the 3GPP standards body caused a new base station class to be defined as well as specific RF performance requirements to be created. This section provides some of the additional requirements with comparisons to the public NodeB when possible. For the complete listing, you are encouraged to refer to [9]. It is important to note that high-speed vehicles are not part of the Home NodeB requirements. Here when we reference high-speed, we are specifically referring to the 120 km/hr and high-speed train requirements.

3.5.1 Base Station Classes

Four base station classes are defined: Wide Area, Medium Range, Local Area, and Home Base Station. The following list further defines them:

- Wide Area Base Stations are characterized by requirements derived from macrocell scenarios with a BS-to-UE minimum coupling loss equal to 70 dB.
- Medium Range Base Stations are characterized by requirements derived from microcell scenarios with a BS-to-UE minimum coupling loss equal to 53 dB.
- Local Area Base Stations are characterized by requirements derived from picocell scenarios with a BS-to-UE minimum coupling loss equals to 45 dB.
- Home Base Stations are characterized by requirements derived from femtocell scenarios.

3.5.2 Frequency Bands

Deployment of femtocells is allowed in any frequency band; this is left up to the service provider. Table 3-7 lists the supported frequency bands.

Please note the nominal channel spacing is still 5 Mhz with a channel raster of 200 KHz for all bands.

Operating Band	UL Frequencies UE Transmit, NodeB Receive	DL Frequencies UE Receive, NodeB Transmit
I	1920–1980 MHz	2110–2170 MHz
II	1850–1910 MHz	1930–1990 MHz
III	1710–1785 MHz	1805–1880 MHz
IV	1710–1755 MHz	2110–2155 MHz
V	824–849 MHz	869–894 MHz
VI	830–840 MHz	875–885 MHz
VII	2500–2570 MHz	2620–2690 MHz
VIII	880–915 MHz	925–960 MHz
IX	1749.9–1784.9 MHz	1844.9–1879.9 MHz
X	1710–1770 MHz	2110–2170 MHz
XI	1427.9–1447.9 MHz	1475.9–1495.9 MHz
XII	698–716 MHz	728–746 MHz
XIII	777–787 MHz	746–756 MHz
XIV	788–798 MHz	758–768 MHz

Table 3-7 Supported Frequency Bands

3.5.3 Base Station Output Power

Maximum output power, *Pmax*, of the base station is the mean power level per carrier measured at the antenna connector in specified reference condition. Table 3-8 provides the rated output power limitations.

BS Class	PRAT
Wide Area BS	- (none)
Medium Range BS	$\leq$ +38 dBm
Local Area BS	$\leq$ + 24 dBm
Home BS	$\leq$ + 20 dBm (without transmit diversity or MIMO) $\leq$ + 17 dBm (with transmit diversity or MIMO)

Table 3-8 Maximum Output Power

TABLE 3-9 Frequency Error Requirements

BS class	Accuracy
Wide Area BS	±0.05 ppm
Medium Range BS	±0.1 ppm
Local Area BS	±0.1 ppm
Home BS	±0.25 ppm

3.5.4 Frequency Error

The modulated carrier frequency of the BS shall be accurate to within the accuracy range given in Table 3-9 observed over a period of one timeslot.

3.5.5 Home Base Station Output Power for Adjacent Channel Protection

The Home BS shall be capable of adjusting the transmitter output power to minimize the interference level on the adjacent channels licensed to other operators in the same geographical area while optimizing the Home BS coverage. The output power, *Pout*, of the Home BS shall be as specified under the following input conditions (the numerical values are provided in Table 3-10):

- CPICH Êc (dBm) is the code power of the Primary CPICH on one of the adjacent channels present at the Home BS antenna connector for the CPICH received on the adjacent channels.
- Ioh (dBm) is the total received power density, including signals and interference but excluding the own Home BS signal, present at the Home BS antenna connector on the Home BS operating channel.

Input Conditions	Output Power, Pout (Without Transmit Diversity or MIMO)	Output Power, Pout (with Transmit Diversity or MIMO)
Ioh > CPICH Êc + 43 dB And CPICH Êc ≥ −105 dBm	≤ 10 dBm	≤ 7 dBm
Ioh ≤ CPICH Êc + 43 dB and CPICH Êc ≥ −105 dBm	≤ max(8 dBm, min(20 dBm, CPICH Êc + 100 dB))	≤ max(5 dBm, min(17 dBm, CPICH Êc + 97 dB))

TABLE 3-10 Output Power for Adjacent Channels

BS Class	Reference Measurement Channel Data Rate	BS Reference Sensitivity Level (dBm)	BER
Wide Area BS	12.2 Kbps	–121	BER < 0.001
Medium Range BS	12.2 Kbps	–111	BER < 0.001
Local Area/ Home BS	12.2 Kbps	–107	BER < 0.001

Table 3-11 Reference Sensitivity Level

In case both adjacent channels are licensed to other operators, the most stringent requirement shall apply for *Pout*. In case the Home BS's operating channel and both adjacent channels are licensed to the same operator, the requirements of this clause do not apply.

3.5.6 Reference Sensitivity Level

The reference sensitivity level is the minimum mean power received at the antenna connector at which the Bit Error Ratio (BER) shall not exceed the specific value indicated in Table 3-11.

3.6 Implementation Issues

Drawing on all of the preceding information, we provide a brief listing of the issues manufacturers of not only the NodeB but also the UE would need to be concerned about when producing a femtocell system. The items to be discussed are

- Timing accuracy
- Power consumption
- Simplified feature set
- Integration with other access technologies

One of the biggest items to be carefully considered is in the area of timing accuracy. As discussed, a new base station class has been identified for the femtocell of the Home NodeB application. The frequency accuracy has been dramatically relaxed from 0.1 ppm to 0.25 ppm. This translates into having a larger uncertainty regarding the actual received frequency. First from the UE perspective, since only hard handoffs are supported in this system, it must be able to adapt from an accurate environment to a less accurate one, when considering moving into the femtocell area. In this case larger residual frequency offset will be realized. The opposite direction is also possible, when moving from a relaxed environment to one that is strict. Second from the femtocell perspective, it must be able to satisfy this requirement of 0.25 ppm over a wide variety of conditions such as aging, temperature, etc. These all lead to architectural design tradeoffs between performance and cost to the end user.

Another item consists of power consumption. Even though this consumer product equipment will be plugged into the power outlet, the additional power consumption the end user will see should be kept to a minimum. This will in turn not cause a large increase in the electric bill. It is imperative not to over-design the femtocell functionality if significant signal processing and more expensive chipsets are to be avoided. We point this out because we expect this femtocell to always be powered on much as cordless telephones operate today.

The femtocell capability should be a simplified version of a public NodeB. This consideration applies to capacity, services provided, administration and maintenance, etc. The initial deployments have converged to supporting a maximum of four users, but going forward we can easily see this value increase. However, the concern is whether the broadband services will accommodate this increase and not become the bottleneck in the architecture. In viewing the progress of the 3GPP performance requirements, one can see there is a need to increase the signal processing capabilities of the receivers. First a conventional RAKE receiver was used in establishing performance targets; shortly thereafter the chip level equalizer (Type 2) was shown to provide improvement in performance. Moreover, interference suppression or interference-aware receivers (Type 2i) produced additional gains in the link budget.

These receiver techniques have assumed a single receive antenna; however, it is well known that antenna diversity techniques improve performance. Hence there exists a dual receive antenna diversity receiver with RAKE (Type 1) and one with equalizers (Type 3). Many WLAN access points already employ multiple antenna techniques in their equipment and at a reasonable price point. It is expected that femtocells will employ these techniques as well as MIMO, not only to deliver high-quality performance but to use this as a baseline where continued improvements can be made.

We will switch our attention to the transmitter now and discuss how significantly lower power is needed for the femtocells when compared to the macro- or microcell deployment. This stems from the fact that not only the cell size has significantly decreased, but also the wireless environment is also different. This means lower power amplifiers can be used and potentially less costly transmit architecture. A final point we would like to mention here is that simplifications can lead to reduced factory calibration time, which should also be considered in the overall cost of the femtocell.

Another item for concern is over integration of other services. For example, a home user can potentially own a cable modem, an access point with built-in router capability, and a cordless phone. Now the home user is expected to also find a place for the femtocell among the devices we have mentioned. It would behoove the manufacturer to integrate other functionality into the equipment to make it more attractive to the end user. For example, Figure 3-13 indicates some possible approaches. First include the WLAN capability into the femtocell. It is understood there may be concerns about this inclusion, especially from the service providers, but in the end peaceful coexistence will prevail and result in a successful deployment. It would also be beneficial to include router capabilities, since desktops may not include the wireless capability. Since VoIP deployments continue to rise, the phone connection (RJ11) to the home is required, at least until all traffic is carried over the wireless medium.

Since many homes are connected via cable, possible integration of the cable modem would be extremely beneficial, although costly. These trade-offs should be discussed in connection with a survey of various home users. Moreover, USB interfaces or SD card interfaces would be good for various storage or multimedia applications. GPS is also shown in order to support the femtocell in registering with the service provider. Other techniques can be used in conjunction with the coordinates, such as collecting the macrocell BS IDs and reporting this list back to the gateway, reporting back an IP address from the broadband service provider, etc. In fact, the most reliable approach would be to have some combination of all of the approaches. Also integrating the functionality should save on overall power consumption. Finally, the Power Line Communication (PLC) protocol has advanced considerably in the last ten years and offers a viable alternative for distribution, if needed.

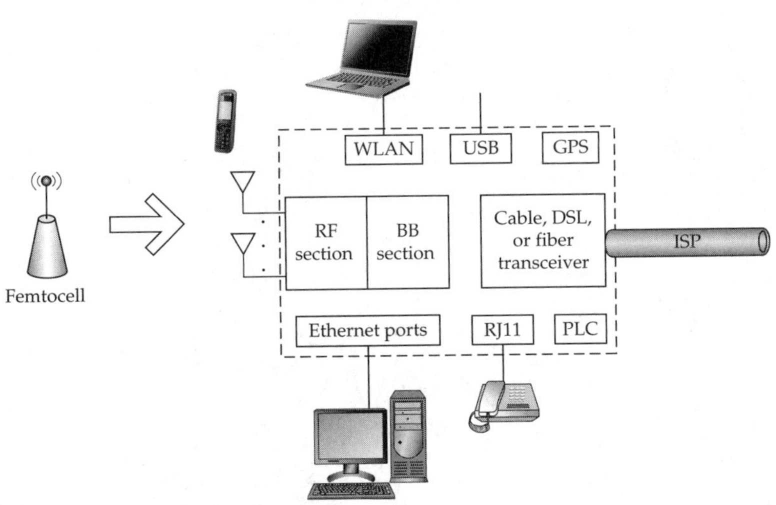

FIGURE 3-13 Example features supported by the femtocell

References

[1] D. Akerberg, "Properties of a TDMA PICO Cellular Office Communication System," Proceedings of IEEE Globecom Conference, December 1988, pp. 1343–1349.

[2] COST Action 231, "Digital mobile radio towards future generation systems," final report, European Commission, Brussels, 1999.

[3] S. S. Ghassemzadeh, L.J. Greenstein, A. Kavcic, T. Sveinsson, and V. Tarokh, "UWB Indoor Path Loss Model for Residential and Commercial Buildings," IEEE Vehicular Technology Conference 2003, pp. 3115–3119.

[4] S. Phaiboon, "An Empirically Based Path Loss Model for Indoor Wireless Channels in Laboratory Building," IEEE Proceedings of TENCON, 2002, pp. 1020–1023.

[5] M. Lott and I. Forkel, "A Multi-Wall-and-Floor Model for Indoor Radio Propagation," IEEE Vehicular Technology Conference 2001, pp. 464-468.

[6] T. Chrysikos, G. Georgopoulos, and S. Kotsopoulos, "Site-Specific Validation of ITU Indoor Path Loss Model at 2.4GHz," IEEE Symposium on a World of Wireless, Mobile and Multimedia Networks, 2009, pp. 1–6.

[7] S. R. Saunders, *Antennas and Propagation for Wireless Communication Systems*, J. Wiley & Sons, 1999.

[8] J. Boccuzzi, *Signal Processing for Wireless Communications*, McGraw-Hill, 2008.

[9] 3GPP Technical Specification (TS) 25.104, Base Station (BS) radio transmission and reception (FDD), Version 9.4.0, 2010.

CHAPTER 4

Femtocell Architectures

This chapter describes details on Femtocell architectures for the various mobile wireless technologies in use today. Discussed are the major components, responsibilities, and their interfaces. The industry accepted reference model is described. Also included are device registration and call flow scenarios for each of the different architectures presented.

4.1 Femtocell Reference Interfaces

Femtocell technology can be used for most if not all mobile wireless technologies, including CDMA, GSM, UMTS, WiMax, and LTE. As femtocell products and services have developed, the implementation details have varied considerably. In order to promote a more consistent approach to femtocell network designs, the Femto Forum has defined a reference model that is intended to apply to all femtocell networks.

The Femto Forum [1] is a telecommunications industry consortium that is chartered with promoting the deployment of femtocell networks. The Femto Forum works in concert with the various standards bodies, which include 3GPP [2], 3GPP2 [3], and the Broadband Forum [4].

Figure 4-1 shows a generic femtocell network diagram that includes the main components and identifies the interface references defined by the Femto Forum.

The following is a description of the main components that can exist in a femtocell network and the associated Femto Forum–defined reference points.

- The femto access point (FAP) is a hardware device located at the customer's premise that interfaces with mobile devices over-the air radio interface. The FAP appears to a mobile device as an outdoor macrocell; however, it emits significantly

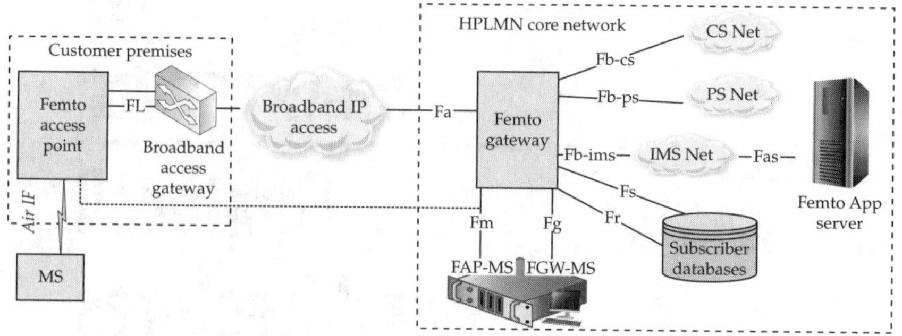

Figure 4-1 Femto Forum reference points

less power. The FAP interfaces with the core mobile network via a broadband interface such as fiber, cable, or DSL. FAPs can support a varying number of mobile devices, typically ranging from four for residential service to eight for small businesses. Using a FAP femtocell to gain access to the mobile network dramatically improves indoor coverage and bandwidth availability for mobile devices. Some FAPs use an Fl interface to control parameters on the broadband access gateway. If a FAP has an integrated access gateway, then the Fl interface is implemented internal to the FAP.

- The femto gateway (FGW) interfaces with the FAP over the broadband access network. This is shown as reference point Fa in Figure 4-1. The femto gateway performs signaling protocol conversion and in some implementations bearer (voice, video…) channel conversions. The femto gateway also performs the function of a security gateway by protecting the mobile network operator (MNO) from attack attempts over the public broadband access at reference point Fa. The FGW interfaces with the different MNO network segments that have been defined as reference points. Fb-cs is the reference point between the FGW and the MNO's circuit-switch network for transporting real-time applications such as voice and video. The Fb-ps is the reference point between the FGW and the MNO's packet-switch network for routing user data such as text messages and e-mail. The Fb-ims interface is used by the FGW to communicate to the core IMS network.
- The subscriber databases are used to store customer provisioning information needed to offer service—information such as the FAP identity and associated settings for provisioning the FAP. The Femto Gateway gets access to the subscribers databases via the Fs and Fr interfaces.

- The femto management system supports two reference points, Fm and Fg. The Fm interface is used to manage the FAP device. The FAP provisioning protocol operates over the IP network between the FAP management system (FAP-MS) and the FAP. FAP vendors and service providers' management protocols vary; however, they generally provide mechanisms for provisioning as well as reporting of faults. The FAP-MS would be expected to manage tens of thousands of FAP devices. The Fg interface is used to manage the femto gateway; its functions include signaling protocol settings, traffic management settings, codec provisioning options, and fault and alarm processing parameters. The femto gateway management system (FGW-MS) implements the Fg interface and should be able to manage multiple femto gateway devices. FGW vendors and service providers' management protocols vary; however, they generally provide mechanisms for provisioning as well as reporting of faults.
- The femto application server interfaces to an IMS network at the Fas reference point.

4.2 IMS Femtocell Architecture

IMS-based femtocell architectures leverage the benefits of an IMS network, which include excellent scalability and an infrastructure to build advanced services. Chapter 9 provides details of the IMS components along with registration and call flow scenarios. In an IMS femtocell architecture one of the key aspects is where the SIP user agent is placed. Figure 4-2 shows a SIP user agent residing in the femto access point (FAP). When a mobile station is in the proximity of the femtocell, it communicates with the FAP using the standard air interface just as it does with a macrocell. The SIP user agent in the FAP registers and places SIP sessions for services on behalf of the mobile station utilizing the IMS network via the femtocell gateway. The femto gateway interfaces with the mobile core network to establish multimedia connections between subscribers attached in a femtocell and subscribers attached to the mobile core network.

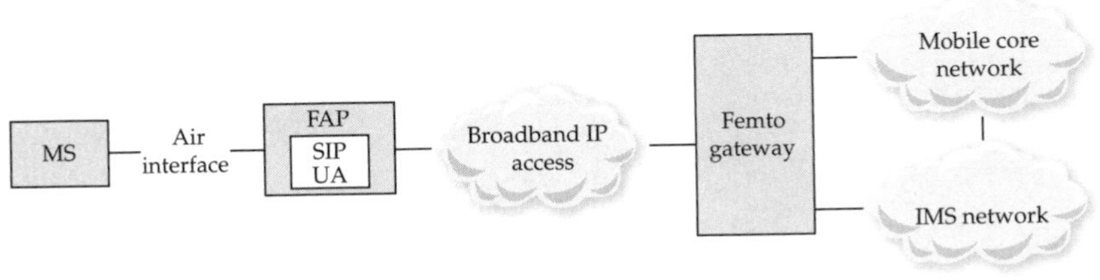

FIGURE 4-2 Femto FAP SIP UA

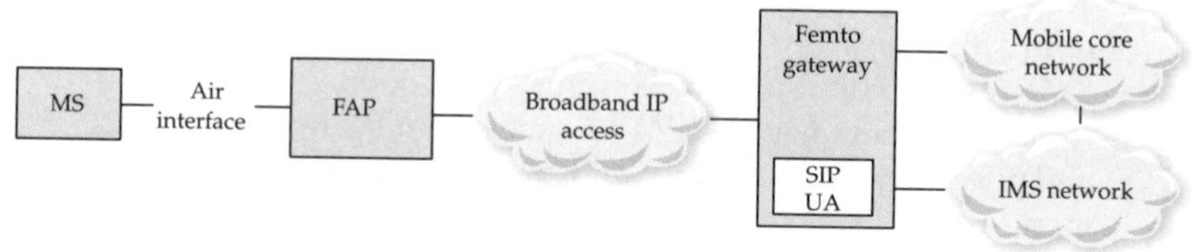

FIGURE 4-3 Femto gateway SIP UA

Communication between the FAP and the femto gateway traverses a public broadband access IP network such as cable, DSL, or FTTH (Fiber to the Home). Since this interface is a public IP network, a security protocol such as IPsec is used to encapsulate the SIP control signaling and RTP. A FAP that uses a SIP user agent is being used for CDMA femtocells. The section "CDMA2000 Femtocells" provides further details on this type of approach.

An alternative to having the SIP user agent reside in the FAP is for it to be implemented in the femto gateway (see Figure 4-3). In this approach the FAP encapsulates the control plane signaling received from the mobile station into an IPsec tunnel. Real-time voice and video signals are encoded using RTP at the FA. The RTP can also be encapsulated into an IPsec tunnel between the FAP and the femto gateway for added security preventing eavesdropping.

4.2.1 CDMA2000 Femtocells

As of the writing of this book, 3GPP2 is defining a CDMA2000 Femtocell Standard [5] that takes the approach of locating the SIP user agent in the FAP. CDMA2000 femtocell deployments to date that use either 1XRTT or EVDO are supported using an IMS approach also with a SIP UA implemented in the FAP. Figure 4-4 is a diagram based

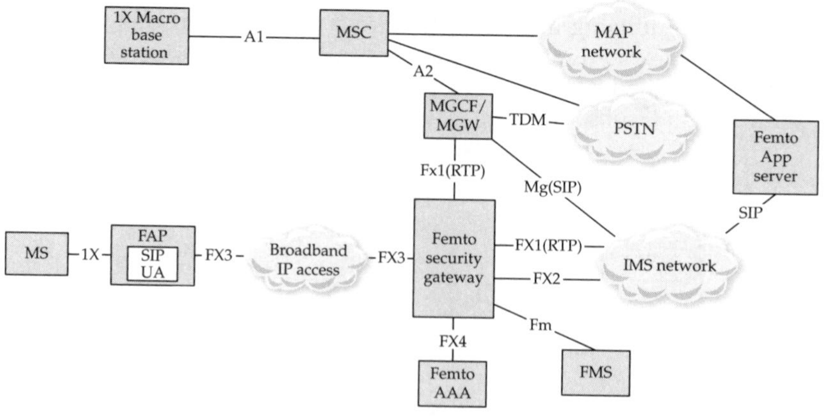

FIGURE 4-4 CDMA2000 femtocell high-level architecture

on the 3GPP2 CDMA2000 femtocell architecture model. The FAP interfaces with the mobile station on the air interface using the standard CDMA2000 protocol stack. Table 4-1 shows a few of the 3GPP2 air interface related standards, some if not all of which a FAP would need to support in order to properly interface with CDMA2000 mobile station.

The main components of the CDMA2000 architecture follow:

- **MS** The mobile station is the same CDMA2000 phone device used to interface with macro base stations.
- **The femto access point (FAP)** This is the device in the femtocell subscriber's home that supports the air interface on one side and an IMS interface on the other. The SIP user agent is implemented in the FAP for registering and signaling to the IMS core network. The FAP accesses the IMS core network using a broadband interface technology such as cable, DSL, or FTTH.
- **Femto security gateway (FSG)** As the name implies, this device offers a secure IP connection between the IMS core network and the FAP via an unsecure broadband access network. This is achieved using IPsec tunnels for control signaling and media between the FSG and the FAP. The FSG can potentially offer value-added services such as transcoding of RTP media streams, allowing for greater interoperability. The FSG interfaces with the femto AAA server for authenticating FAPs. The FSG provides access for the FMS in order to autoconfigure the FAPs.
- **The femto AAA server** This is used for authenticating the FAPs. Once a FAP has been authorized, the AAA server shares security policy data with the FSG to allow IPsec tunnels to be established between the FAP and the FSG.

3GPP2 Specification	Description
C.S.0001	CDMA2000 Introduction
C.S.0002	CDMA2000 Physical Layer
C.S.0003	CDMA2000 MAC Layer
C.S.0004	CDMA2000 Layer2 MAC
C.S.0005	CDMA2000 Layer3
C.S.0010	Base Station minimum standard
C.S.0014	Enhanced Variable Rate Codec (EVRC)
C.S.0015	Short message service
C.S.0016	Over-the-air service provisioning

Table 4-1 CDMA2000 Air Interface–Related Specifications

- **The femto management system (FMS)** This is responsible for configuring and managing the various femtocell components, especially autoconfiguring the FAPs. A design goal of femtocells is to be plug-and-play ready. Having femtocells autoconfigure allows wireless service providers the luxury of having the end customer install the FAP directly without requiring a costly service technician.
- **The femto application server** This is used to support interworking functions between the IMS core network and the mobile carrier's MAP network. The MAP network consists of network elements that include HLR (home location register), MC (message center), and MSC (mobile switching center), to name a few.

The CDMA2000 Femtocell Standard [6] has defined the following reference points as shown in Figure 4-4:

- Fx1 is an interface that carries RTP media packets to and from the femtocell gateway. This includes RTP, which traverses an IP network end to end, or RTP traffic that is converted back to TDM using a media gateway device.
- Fx2 is the SIP signaling control interface between the femtocell gateway and the IMS core network.
- Fx3 is the IPsec tunnel used between the FAP and the femtocell security gateway.
- Fx4 is the SIP signaling interface between the femtocell security gateway and the Femtocell AAA server.
- Fm is the interface between the femtocell security gateway and a femtocell management system for autoprovisioning the FAP.

CDMA2000 Femtocell Signaling Protocols

The air interface between the mobile station and the FAP is intended to be the same as the air interface between the mobile station and a macrocell. The femtocell network provides an alternative and enhanced access to the mobile service provider's network via a broadband access network. Figure 4-5 shows the CDMA2000 control plane-related protocol stacks.

The following describes the protocol stacks shown in Figure 4-5:

- The mobile station (MS) interfaces with the FAP over the physical layer air interface the same as it does with a macrocell. As a result, the protocol stack in the mobile station has no changes specifically for femtocell support. The stack consists of physical layer CDMA2000 air interface, a Media Access Control (MAC) layer, a link access control layer (LAC), and the layer 3 signaling layer. The layer 3 signaling layer performs all of

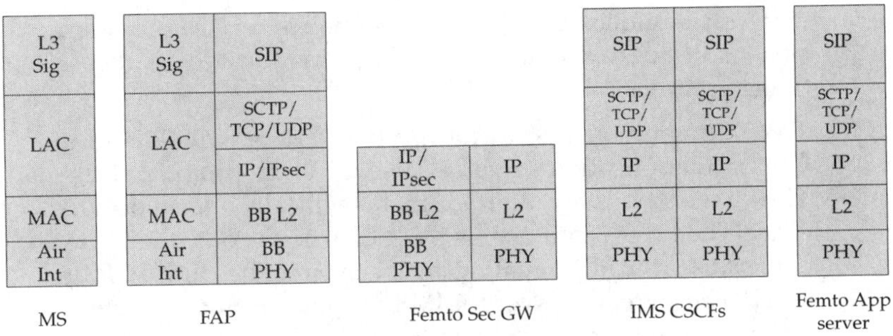

Figure 4-5 CDMA2000 IMS femtocell control protocols

the call establishment procedures defined in the CDMA2000 Layer 3 specification.

- The FAP implements two interfaces, one being the air interface and the other being the broadband access interface. On the air interface it communicates with one or more mobile stations, and as a result the protocol stack looks identical to the MS. On the broadband interface the FAP implements a broadband interface physical layer. This interface can be an Ethernet interface where the FAP gets directly connected to a broadband modem such as a cable, DSL, or FTTH device. The type of broadband physical interface used will dictate the type of broadband layer 2 protocol needed. If it is Ethernet, then the layer 2 protocol is an Ethernet MAC layer protocol.

  The layer 3 protocol consists of IP and IPsec. The IPsec is used to establish secure tunnels for upper-layer signaling. The layer 4 protocol is used to encapsulate the signaling layer, which can be either UDP, TCP, or SCTP. Each of these protocols have advantages and disadvantages. Refer to Chapter 5 for further details. The higher-layer signaling protocol layer uses SIP in order to register and establish sessions with the core IMS network. The FAP will need to convert all layer 3 signaling communication with the MS and map that into SIP messaging to the IMS network. Similarly, any SIP message the FAP receives will need to be mapped into a layer 3 CDMA message to be sent to the MS. This requires the FAP to maintain and coordinate protocol state data between the L3 CDMA air interface and the SIP IMS network interface.

- The femtocell security gateway for signaling control essentially bridges the IPsec tunnel interface on the broadband access side with a pure IP interface on the IMS side. The broadband access requires the use of IPsec because it is an untrusted IP interface, traversing a public IP network. The IMS network is under the domain of the mobile service provider, as a result doesn't require IP traffic to pay the overhead cost of using IPsec.

- The IMS components, including the various CSCFs and a femtocell application server, all use the same protocol stack. They have a physical interface that is typically a form of Ethernet. The layer 3 protocol is IP. The layer 4 protocol used to encapsulate the SIP signaling layer can either be UDP, TCP, or SCTP. The signaling layer for IMS of course is SIP. The femto application server will convert SIP to the Mobile Application Part (MAP) protocol to interface with the mobile core SS7 network.

CDMA2000 Femtocell Media Protocols

This section briefly describes the media-related protocol stacks for a CDMA2000 femtocell network. The media protocols for the air interface between the mobile station and the FAP are the same as for the air interface between the mobile station and a macrocell. Figure 4-6 shows the CDMA2000 media plane–related protocol stacks.

The following describes the protocol stacks shown in Figure 4-6:

- The mobile station (MS) interfaces with the FAP over the physical layer air interface just as it does with a macrocell. A media codec such as voice or video stream is used to encode the media into a CDMA2000 channel over the CDMA2000 air interface.
- The FAP implements two interfaces, one being the air interface and the other being the broadband access interface. On the air interface it communicates with one or more mobile

MS	FAP		Femto Sec GW		MGW
Media encoding	Media encoding	Media encoding			Media encoding
CDMA Air Interface	CDMA Air Interface	RTP/RTCP			RTP/RTCP
		UDP			UDP
		IP/IPsec	IP/IPsec	IP	IP/IPsec
		BB L2	BB L2	L2	BB L2
		BB PHY	BB PHY	PHY	BB PHY

Figure 4-6 CDMA2000 IMS femtocell media protocols

stations, and as a result, the protocol stack looks identical to the MS. On the broadband interface the FAP implements a broadband physical and corresponding layer 2 interface. The layer 3 protocol consists of IP and IPsec, allowing secure tunnels for upper-layer media protocols. UDP is used to encapsulate the upper-layer media protocols. RTP and RTCP are used to carry the real-time encoded media. Media encoding is based on the type of media, which can be voice or video.

- The FSG implements the same protocol stack for both signaling and media as described in the section "CDMA2000 Femtocells" when interfacing with the FAP.
- The IMS network components, such as a media gateway device that terminates media protocols, all use the same protocol stack. They have a physical interface that is typically a form of Ethernet and a layer 2 Ethernet MAC layer protocol. The layer 3 protocol is IP. UDP is used to encapsulate the upper-layer media protocols. RTP and RTCP are used to carry the real-time encoded media. Media encoding is based on the type of media, which can be voice or video.

CDMA2000 Femtocell FAP Registration

This section describes an IMS femtocell FAP registration scenario. In order for the IMS network to know the existence of any FAP, it needs to register itself with the IMS femtocell application server. The FAP typically would initiate the registration procedure any time it boots up, such as in a power-up scenario. Figure 4-7 shows the IMS FAP registration procedure.

A description of each message in Figure 4-7 scenario follows:

1. At power-up the FAP sends a SIP Register Message to its Proxy CSCF. Included in the registration message is the FAP's FQDN (fully qualified domain name). The FQDN is the network name assigned to the FAP that can be mapped to an Internet address as is typically performed by a DNS (Domain Name Server). The FQDN is usually of the format FAPName@ DomainName.
2. The P-CSCF performs IPsec security functionality and forwards the SIP Registration message to the I-CSCF.
3. The I-CSCF sends a Diameter UAR to the HSS to obtain the address of the proper S-CSCF to send the registration to.
4. The HSS replies with a UAA message that contains the proper S-CSCF to progress messaging.
5. The I-CSCF sends the registration to the specified S-CSCF.
6. The S-CSCF checks the Authorization of the FAP with the HSS by sending a diameter MAR message.

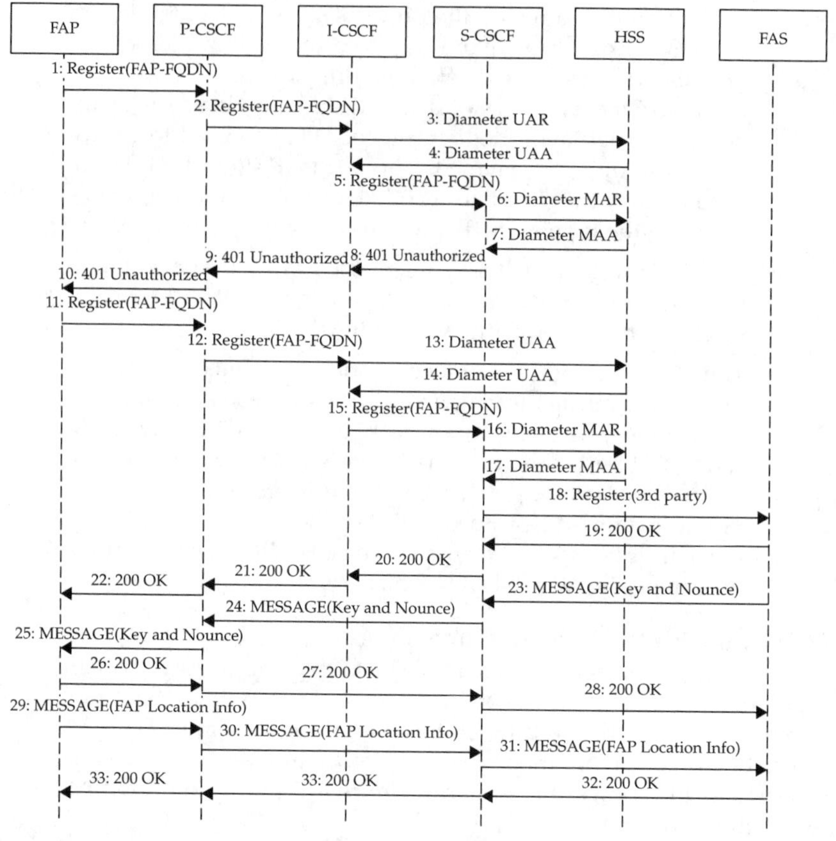

Figure 4-7 FAP IMS registration

7. The HSS checks if the FAP has provided its credentials yet. Since it hasn't provided credentials, it sends a diameter MAA reply indicating that the credentials are not known.

8–10. The S-CSCF sends a challenge request for the FAP's credentials by sending a 401 Unauthorized SIP message. Messages 8–10 are the propagation of the 401 from the S-CSCF to the FAP.

11–17. Messages 11–17 are similar to 1–7 with the FAP reattempting to register, this time specifying its credentials to allow for proper authorization.

18. After the S-CSCF receives confirmation that the FAP is properly authorized, it registers the FAP with the FAS (femtocell application server) by forwarding it the registration message.

19–22. These messages acknowledge the acceptance of the registration from the FAS back to the FAP.

23–25. The FAS sends a SIP MESSAGE containing the initial configuration data that the FAP needs for any future message processing.

26–28. The FAP sends the FAS a 200 OK message acknowledging the SIP message.

29–31. The FAP sends a SIP message to the FAS informing it of its radio neighbor list, including other FAPs and macrocells.

32–33. The FAP sends the FAS a 200 OK message acknowledging the SIP message.

CDMA2000 Femtocell Mobile Station Registration

This section describes an IMS femtocell mobile station registration scenario. After a FAP has successfully registered, it allows one or more mobile stations to register via the FAP. Mobile stations when first turned on search for the closest mobile base station. Based on its signal strength, the mobile station registers itself with the CMDA network. This allows the network to know where the mobile station can be reached in the event a call arrives destined for that mobile station. For a femtocell, a mobile station will pick up the FAP's radio signal as the nearest base station and attempt to register with it. Figure 4-8 shows the IMS FAP registration procedure.

1. When a mobile station detects the FAP as the closest base station based on radio signal strength, it attempts to register by sending a CDMA2000 1X Registration Request.

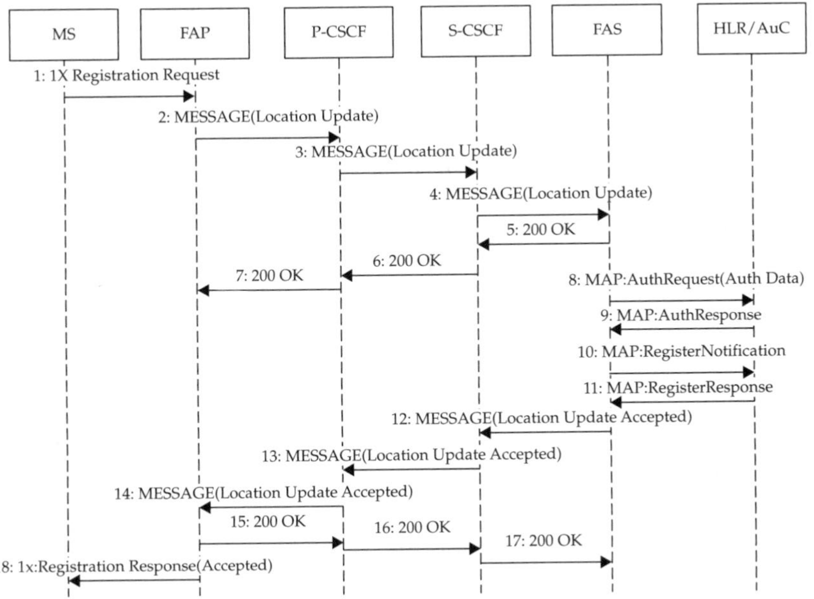

Figure 4-8 Mobile station registration via FAP

2. When the FAP receives the 1X CDMA2000 Registration Request, it then sends a SIP MESSAGE containing a Location Update indication and associated parameters.

3–4. The SIP MESSAGE sent from the FAP is forwarded to the FAS via the P-CSCF and the S-CSCF.

5–7. The FAS acknowledges the SIP MESSAGE with a 200 OK, which gets forwarded back to the FAP.

8. The FAS sends a MAP authentication request that contains the CDMA2000 authentication data provided by the mobile station to the AuC.

9. The AuC authenticates the mobile station and sends a successful MAP Authentication response. If the AuC responded with a failed registration, the FAS would then send a 401 unauthorized SIP message back to the FAP.

10. The FAS sends a MAP Registration Notification to the HLR that contains the FAS as the serving MSC.

11. The HLR replies with a successful Registration Response. If the registration fails, the FAS will send a SIP error code back to the FAP indicating the registration failure.

12–14. The FAS sends a SIP MESSAGE indicating the location update was successful. This message is forwarded from the FAS to the FAP. Included in the message is the mobile directory number for the registering mobile station.

15–17. The FAP acknowledges the MESSAGE by sending a 200 OK. The 200 OK gets forwarded back to the FAS.

18. The FAP sends a CDMA2000 1XRegistration Response indicating a successful registration.

CDMA2000 Femtocell Mobile Call Origination

This section describes an IMS femtocell call origination. After both the FAP and a mobile station have successfully registered, calls can be originated or terminated by the mobile station. Figure 4-9 shows the IMS-based femtocell call origination.

1. When a user wants to place a call, he enters a phone number in his mobile station, which in turn generates a 1X-Origination message sent over the air interface. Included in the message is the called phone number as well as authorization information.
2. The FAP in response to receiving the origination seizes and establishes a channel on the air interface to be used to carry the voice path over the air interface.

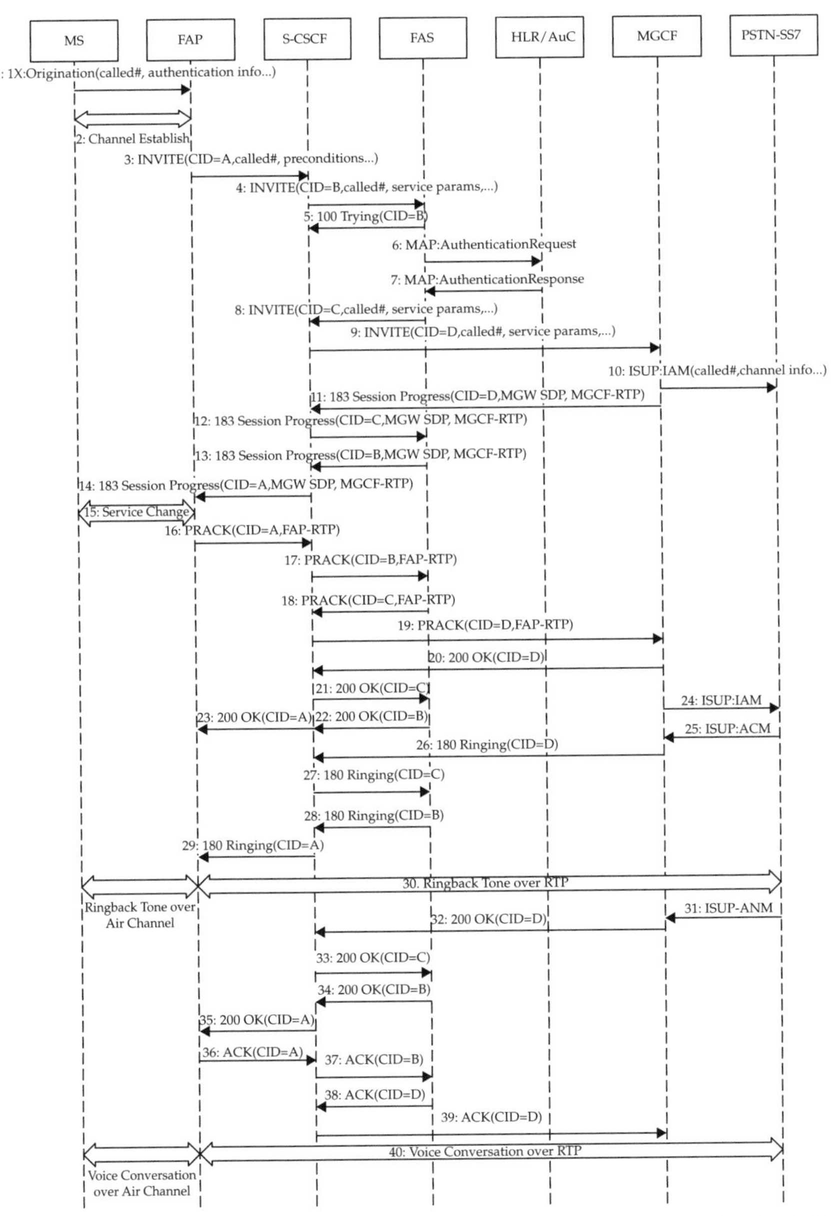

FIGURE 4-9 Femtocell IMS call origination

3. The FAP attempts to originate a call by sending an INVITE message to the CSCFs (P-CSCF, S-CSCF). Included in the INVITE are the called party number, the MAP authentication parameters, and the calling mobile directory number.

4–5. The INVITE is propagated to the FAS along with a 100 trying sent in response.

6–7. The FAS authenticates the mobile stations request with the AuC.

8–11. The FAS establishes a TDM connection via a MGCF (Media Gateway Control Function) by sending an INVITE to the MGCF via the S-CSCF. The INVITE includes the called party number and service parameters. Once a successful path is established through the PSTN SS7 network, a SIP 183 Session Progress message is returned from the MGCF that includes the MG (Media Gateways) SDP.

12–14. In response to receiving a 183 from the MGCF, the FAS propagates a 183 back to the FAP informing it of the SDP of the MGCF.

15. The FAP modifies the air interface channel if necessary based on the SDP channel characteristics received in the 183.

16–19. In response to the 183, the FAP sends a PRACK and includes its SDP to be sent back to the MGCF. The PRACK is propagated back through the FAP to the MGCF.

20–23. The PRACK is acknowledged with a 200 OK propagated from the MGCF back to the FAP.

24. The MGCF sends an ISUP IAM in response to the PRACK received in Step 19.

25. When the far end phone is being alerted, the PSTN responds with an ISUP-ACM to the MGCF.

26–30. A SIP 180 ringing message is propagated from the MGCF back to the FAP indicated that the called party is alerted. Step 30 shows that the ringback tone is provided over the established RTP channel.

31. Eventually the called party answers her phone, which results in an ISUP-ANM message being sent to the MGCF from the SS7 network.

32–35. When the MGCF receives the ISUP:ANM, it in turn sends a 200 OK for the INVITE received in Step 9 indicating that the call has been answered. The 200 OK is propagated back to the FAP.

36–39. The 200 OK is acknowledged with an ACK that is sent from the FAP back to the MGCF.

40. A two-way conversation now takes place over the voice channel on the air interface, the RTP channel over the IP network, and the channel allocated on the PSTN.

Femtocell IMS Call Release

This section describes an IMS femtocell call release. Active calls can be released by either the calling or called party. Figure 4-10 shows a call release initiated from a CDMA2000 mobile station for an active call established on a femtocell access network. A description of each of the messages is provided.

1. When the CDMA2000 Mobile User attempts to end an active call, it results in the mobile station sending a 1X Release message over the air interface.
2. When the FAP receives the 1X Release, it releases the allocated channel associated with the air interface.

3–6. The FAP initiates the SIP session release by sending a BYE message that gets propagated to the MGCF.

7. The MGCF initiates releasing PSTN resources for the call by sending an ISUP release message.

8–11. The session release is confirmed by the MGCF sending a 200 OK for the BYE that is propagated from the MGCF to back to the FAP.

CDMA2000 Femtocell to Macrocell Handoff

This section describes a handoff of an IMS femtocell call to a macrocell. This is a typical scenario where a mobile subscriber is in her home with an active femto call and leaves home during her conversation. As the subscriber moves further away from her femtocell, the mobile station detects the signal power reduction and initiates a handoff procedure. The handoff procedure for the mobile station is the same as a handoff that occurs between macrocells as a phone moves

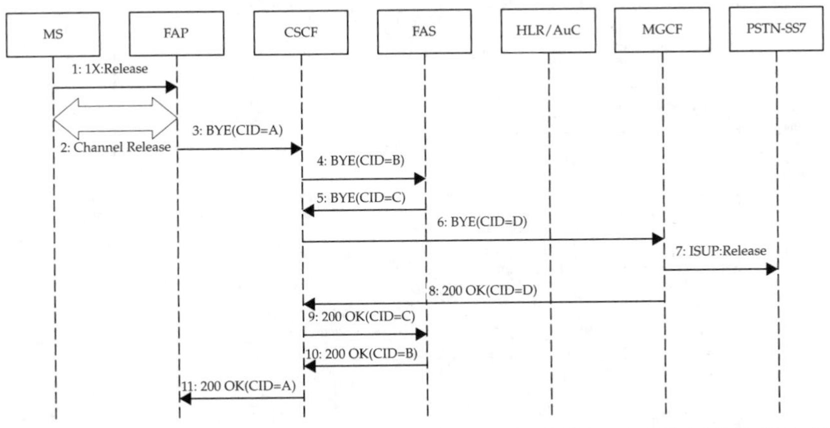

FIGURE 4-10 Femtocell IMS call release

from one macrocell to another. Figure 4-11 shows a femtocell-to-macrocell handoff. A description of the messaging follows.

1. As the mobile station moves further away from the FAP, it detects that it is receiving a better signal from an adjacent macrocell. The MS then indicates that a handoff should be performed by sending a Radio Signal Measurement report to the FAP.

2–3. The FAP verifies that a handoff is required and sends a SIP MESSAGE containing a Handoff Request from the FAP to the FAS via the CSCF.

4–5. The Handoff Request is acknowledged by the FAS with a 200 OK.

6. The FAS attempts to create a channel to the target MSC by sending an INVITE to the MGCF. The MGCF has a circuit switch access to the targeted MSC. The circuit switch access is typically a DS0 channel of a T1 interface. The INVITE contains the SDP of the other endpoint OEP and the ID of the target MSC; it specifies send only in the SDP.

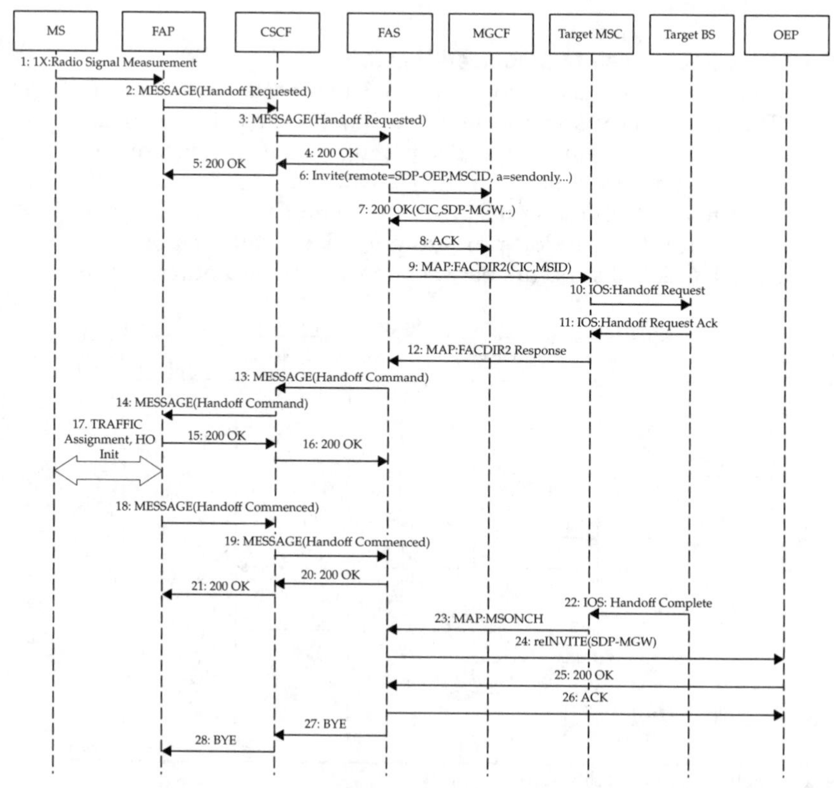

FIGURE 4-11 Femtocell to macro network handoff

7–8. The MGCF accepts the INVITE and allocates a circuit-switch channel to the MSC and provides a 200 OK message back to the FAS. Contained in the 200 OK is the SDP of the MGW channel. Message 8 the FAS sends an ACK for the 200 OK.

9. The FAS sends a MAP FACDIR2 message to the target MSC indicating the MGCF's channel to be used for the handoff.

10. When the target MSC receives the FACDIR2 message, it in turn sends a handoff request message to the target Base Station.

11. The Target BS determines if it can honor the handoff based on the resources requested and accepts the request by sending a Handoff Request ACK back.

12. The Target MSC after receiving the Handoff Request ACK sends a MAP:FACDIR2 response.

13–16. The FAS proceeds with the handoff by sending a SIP message that contains a handoff command to the FAP via the CSCF. The FAP acknowledges the handoff command by sending 200 OK back to the FAS.

17. The FAP instructs the mobile station to hand off its air interface channel from the FAP to the macro base station.

18–21. The FAP informs the FAS that the handoff has occurred by sending a SIP Message containing a handoff commenced indication. The FAS acknowledges this message by sending 200 OK.

22. Once the Target BS detects that the handoff succeeded, it sends an IOS Handoff Complete message to the Target MSC.

23. The Target MSC after receiving the Handoff Complete message sends a MAP:MSONCH to the FAS.

24–26. The FAS sends a reINVITE to the other endpoint providing the SDP of the channel on the Media Gateway that has been allocated for the handoff to the target MSC.

27–28. Since the session between the FAP and the FAS is no longer needed, the FAS sends a SIP BYE to release the original session.

CDMA2000 Macro-to-Femtocell Handoff

This section describes a handoff of a macrocell call to an IMS femtocell call. This is a typical scenario where a mobile subscriber is on the road and arrives home with an active call. The benefits of switching from the macrocell to the femtocell include a better-quality connection with improved bandwidth availability. Also, femtocell billing typically is more favorable than using the macrocell network. As the subscriber moves closer to a femtocell, the mobile station detects that the signal power of the femtocell is significantly stronger than the signal from the macrocell. The handoff procedure for the mobile station is the same as a handoff that occurs between macrocells as a phone moves from one macrocell to another. Figure 4-12 shows a macrocell-to-femtocell handoff. A description of the messaging is provided.

1. At the start a mobile station has an active call between itself and another endpoint (OEP).
2. As the mobile station moves closer to the FAP, it detects that it is receiving a better signal from the FAP compared to the macrocell. The MS then indicates that a handoff should be performed by sending a Radio Signal Measurement report to the macro base station.

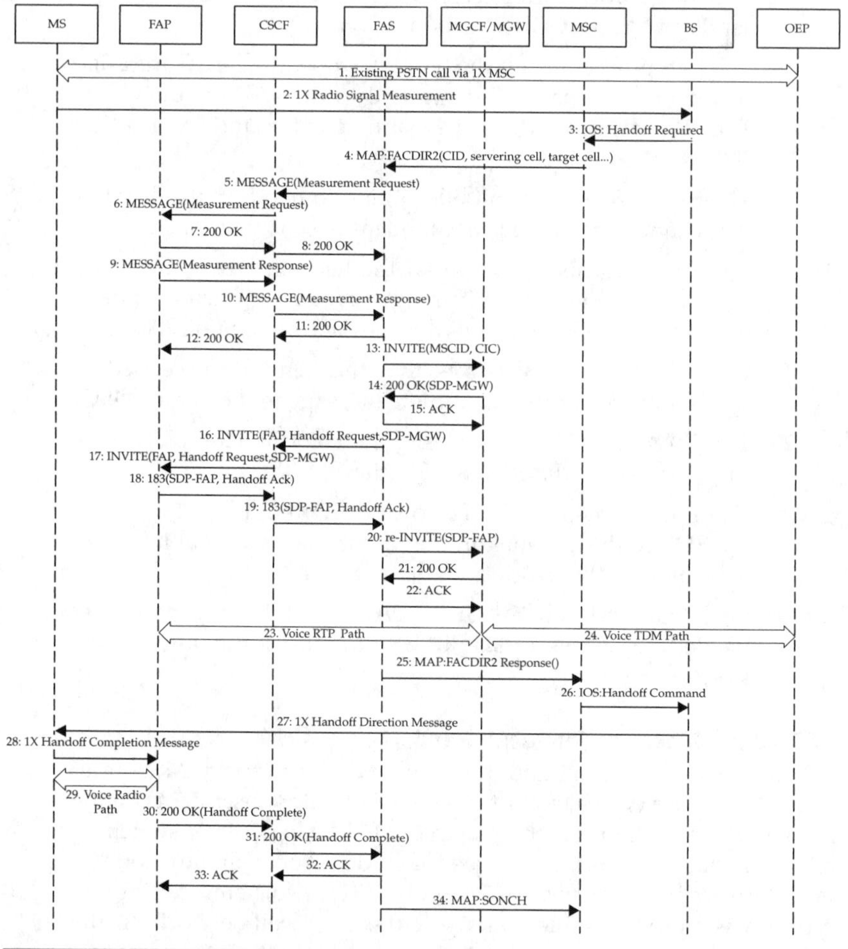

FIGURE 4-12 CDMA2000 IMS macrocell-to-femtocell handoff

3. The macro base station determines that a handoff should occur to the FAP and sends a IOS Handoff Request message to the macro MSC.

4. The macro MSC determines from the MSCID that handoff should be processed by the FAS. The MSC sends a MAP FACDIR2 message to the FAS indicating that a handoff should take place.

5–12. The FAS validates the signal strength a FAP is receiving by sending a SIP MESSAGE that contains a measurement request. The FAS could send this request to multiple FAPs in order to determine which FAP has a better signal from the mobile station. The FAP responds with a measurement response indicating the signal strength being received by the FAP.

13. The FAS attempts to create a channel to the OEP by sending an INVITE to the MGCF. The MGCF has a circuit-switch access to the OEP. The circuit-switch access is typically a DS0 channel of a T1 interface.

14–15. The MGCF accepts the INVITE, allocates a circuit-switch channel to the MSC, and provides a 200 OK message back to the FAS. Contained in the 200 OK is the SDP of the MGW channel. In message 15 the FAS sends an ACK for the 200 OK.

16–17. The FAS initiates a handoff with the FAP by sending an INVITE with a handoff request that includes the SDP of the MGCF.

18–19. The FAP accepts the INVITE by sending a 183 session progress back toward the FAS that includes a Handoff ACK indication and the SDP of the FAP.

20–22. The FAS updates the MGCF with the FAPs SDP by sending a re-INVITE to the MGCF. The MGCF accepts the FAP's SDP and replies to the FAP with a 200 OK.

23–24. At this point a voice path has been established between the FAP and the MGCF using RTP over the femtocell IP access network. The MGCF has a circuit-switch connection allocated to the OEP.

25. The FAS indicates that the handoff is ready to occur by replying to the MSC with a MAP:FACDIR2.

26–27. The Macro MSC instructs the mobile station via the base station to execute a handoff.

28. The MSC informs the FAP that the handoff has completed by sending a CDMA2000 1X Handoff Completion Message.

29. The FAP establishes a voice channel with the mobile station.

30–33. The FAP informs the FAS that the handoff session is active by sending a 200 OK to the INVITE in Step 16 containing a Handoff Complete. The FAS sends an ACK in response.

34. The FAS informs the MSC that the handoff is complete by sending a MAP SONCH message.

4.3 UMTS Femtocell Architecture

Universal Mobile Telecommunications Service (UMTS) is a third-generation (3G) wireless mobile technology that evolved from GSM. It is very widely deployed and as such is one of the first wireless technologies to be standardized for femtocells. The 3GPP standards body defined the UMTS Home NodeB (HNB) architecture [7] in 2009. Figure 4-13 shows a high-level architecture diagram of the UMTS femtocell network. This architecture is consistent with the generic Femto Forum architecture and is described in the text that follows.

The mobile phone, also known as the user equipment (UE), interfaces with a Home NodeB (HNB) over the air interface. The Home NodeB is a device that resides in the customer's home that is analogous to the Femto Forum's FAP. The HNB's name stems from the fact that it is a miniature UMTS NodeB providing wireless coverage in the customer's home. The air interface between the UE and the HNB is defined as the Uu reference point. The HNB interfaces to the mobile network operator (MNO) over a broadband network by interfacing with a broadband access gateway.

The mobile operator's core network obtains access to the HNB via the broadband access network at the Iu-h reference point. At the Iu-h reference point a mobile service provider uses a security gateway to protect the core network against attacks. On the trusted side of the security gateway resides the Home NodeB Gateway (HNB GW). The HNB GW is analogous to the Femto Forum's FAP GW, which is responsible for interfacing with the mobile operator's core network's Iu interface. The Iu interface consists of the Iu-cs for circuit-switched traffic and the Iu-ps for packet-switched traffic.

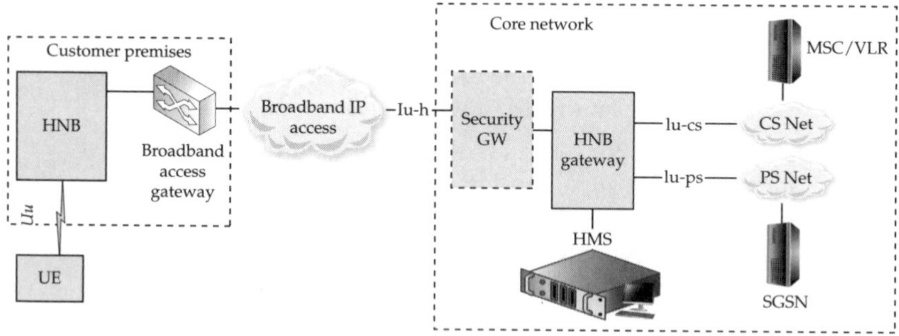

Figure 4-13 3GPP Iu-h femtocell reference model

4.3.1 UMTS Femtocell Signaling Protocols

Figure 4-14 shows the Iuh control protocol stack used by femtocell networking equipment. The user equipment (UE) stack is the same stack it uses to interface with macrocell NodeB, since the UE sees the HNB equivalent to a macrocell NodeB. At the physical layer is the radio frequency (RF) interface, also referred to as the air interface. The higher-layer protocols gain access to the RF interface via the Media Access Control (MAC) layer. Above the MAC layer the UE's higher-layer control protocols use the Radio Link Control (RLC) layer to gain access to various radio channels to communicate to the mobile network. The RLC provides error-free delivery of messages through mechanisms of error detection, message acknowledgments, message sequence number, and message retransmissions. The Radio Resource Control (RRC), as the name implies, is responsible for managing the radio resources that are allocated for user equipment. The RRC is responsible for establishing and releasing radio channels, including signaling and bearer channels. Radio signal measurements are invoked and reported by the RRC, allowing it to also perform mobility management. The mobility management includes determining if a call should be handed over, cell reselection, and routing updates.

The Radio Access Network Application Part (RANAP) protocol is responsible for radio access bearer (RAB) management. RABs are bearer channels over radio interfaces used for transporting media, including voice and data. The RANAP RAB management includes signaling protocol messages for establishing, releasing, and modifying radio bearer channels. The RANAP User Adaptation (RUA) protocol has been specifically designed for carrying RANAP messages between an HNB and an HNB gateway. The RANAP messages are carried transparently without any alteration. Notice in Figure 4-14 that the RUA is strictly implemented on the HNB and the HNB gateway.

The RANAP protocol is an existing protocol that is used today in UMTS networks for establishing and releasing calls. The Home NodeB Application Protocol (HNBAP) [10] was defined by 3GPP specifically

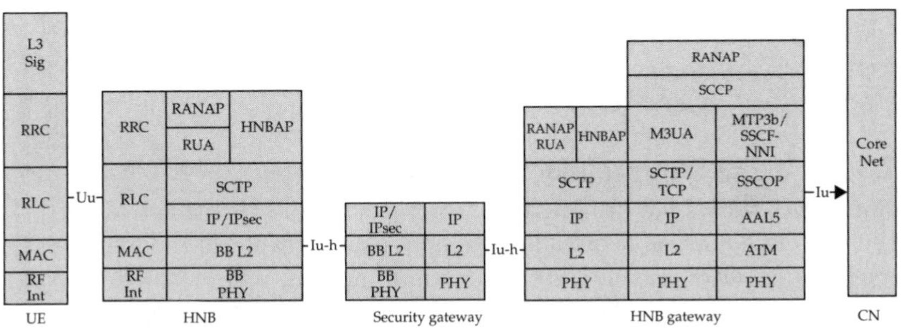

FIGURE 4-14 Iu-h femtocell control protocols

for Home NodeB applications. The main purpose of this protocol is for HNB registrations with the HNB gateway. Also defined in the HBBAP protocol is the registration of the UE in an HNB network. Both the RUA and the HNBAP protocols operate over the Stream Control Transmission Protocol (SCTP). SCTP, similar to TCP, is a transport-layer protocol providing reliable delivery of messages between the HNB and the HNB gateway. At the IP layer, IPsec is used for security purposes, since the communications between the HNG and the HNB gateway are over an untrusted IP network. The broadband layer 2 (BB L2) and physical layer (BB Phy) depend upon the type of broadband access being used by the subscriber, such as Fiber to the Home (FTTH), cable, or DSL.

The HNB gateway is responsible for interfacing with the HNB on the broadband access side via a security gateway. On other interfaces the HNG gateway interfaces with mobile service provider's core network over the Iu interface. Iu interfaces can be based on an IP network or the conventional ATM network. Figure 4-14 shows an HNG gateway that supports both, whereby one stack is ATM based and the other is IP based. As wireless networks evolve, the trend is to move toward an all-IP-based network. Both stacks at the highest layer on the HNB gateway use the RANAP protocol for managing radio access bearer channels. RANAP uses the services of the SS7 Signaling Connection Control Part (SCCP) that is used for routing, flow control, and error correction of RANAP messages.

Below the SCCP layer is where the ATM and IP stacks diverge. The IP stack uses the service of the Message Transfer Part 3 User Adaptation (M3UA) for transporting the SCCP messaging over the SCTP IP stack. M3UA in general is used for transferring SS7 MTP3 User Part messages over SCCP, which includes RANAP, ISDN User Part (ISUP), and Telephone Users Part (TUP). On the ATM stack SCCP uses the services of the Layer 3 Broadband Message Transfer Part (MTP3b), which functions as the MTP3 layer with some changes to interface to an ATM subnetwork. The MTP3B protocol uses the services of the Service-Specific Coordination Function at the Network Node Interface (SCCF-NNI) and the Service-Specific Connection-Oriented Protocol (SSCOP) to establish and release the control channels. SSCOP is similar to SCTP and TCP in that it performs the function of a transport protocol. SSCOP has been designed specifically for ATM-based networks and usually uses the services of the ATM Adaptation Layer 5 (AAL5) layer for sending and receiving SSCOP packets.

4.3.2 UMTS Femtocell Media Protocols

Figure 4-15 shows the Iu-h media protocol stack used by the femtocell networking equipment. Just like the control protocol stack, the media stack is the same for the UE when communicating with a macrocell node. Above the MAC layer the UE uses media encoding for applications such as speech. The HNB routes the media encoded stream from

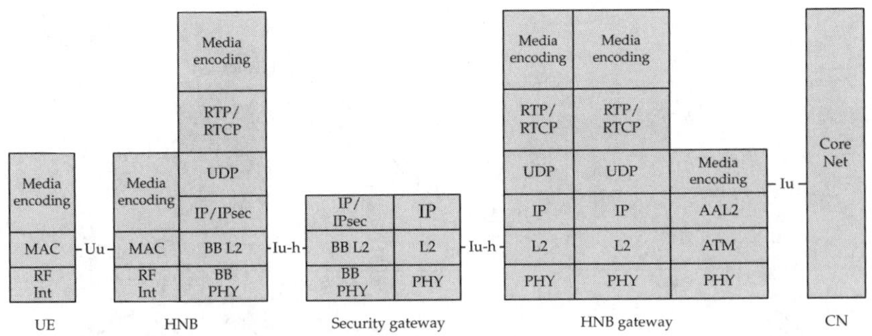

FIGURE 4-15 Iu-h femtocell media protocols

the air interface into the broadband access interface stack. RTP/RTCP is used on top of UDP to transport the real-time media stream from the HNB to the HNB gateway. At the IP layer for increased security IPsec can be used. Because IPsec requires significant network resources, service providers may opt to not use it for media transport. At the HNB gateway the Media Stream is routed to an Iu interface toward the mobile operator's core network. Again two stack options here are either IP or ATM. For IP the lower-layer protocols remain the same as used at the Iuh, which are RTP/RTCP. For an ATM-based stack, the lower layer used is the ATM Adaptation Layer 2 (AAL2). AAL2 is the adaptation layer used by ATM for carrying real-time traffic with high QoS.

4.3.3 UMTS Femtocell Device Responsibilities

Figure 4-16 shows a high-level diagram of the main components for a UMTS femtocell network. It identifies the functional responsibilities of each of these components.

4.3.4 HNB Registration

This section describes the HNB registration scenario illustrated in Figure 4-17. In order for the UMTS core network to know the existence of any HNB, it needs to first register itself. The HNB typically would initiate the registration procedure any time it powers up.

1. When the HNB is first powered up, it goes through an initialization process using factory set options. On power-up after the HNB has already been through a successful registration, it will use the settings it obtained from the network management system.
2. The HNB initiates establishing a IPsec tunnel to the security gateway.
3. After the IPsec tunnel has been established, the HNB initiates establishing a SCTP session to a well-known port that the HNB gateway is listening on.

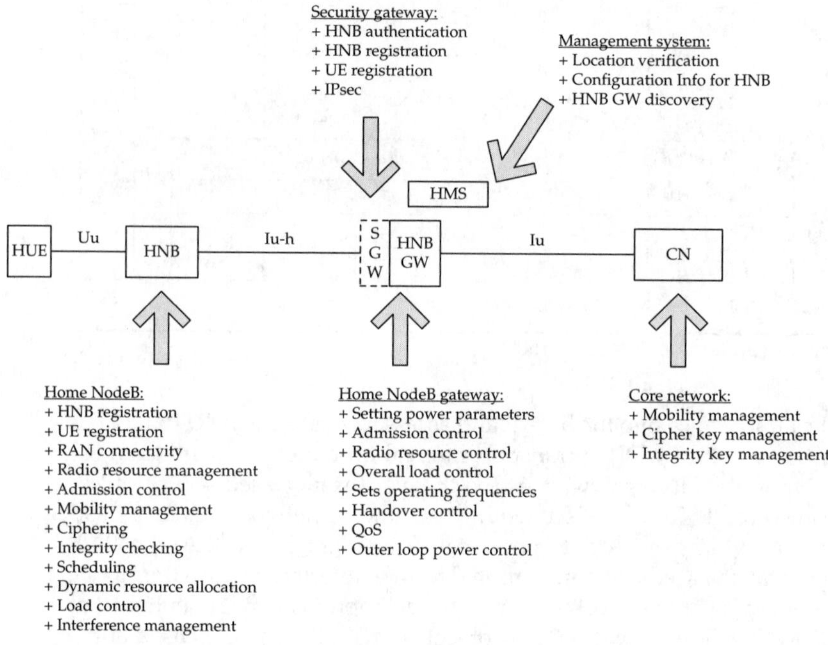

Figure 4-16 Iu-h femtocell device responsibilities

4. After the SCTP channel is established, the HNB sends a registration request message. Contained in the registration request message is the HNB location information it obtains either from GPS or from the macrocell network.
5. The HNB gateway processes the registration request and determines if the HNB is allowed to operate given its location and an interference calculation. After all parameters have

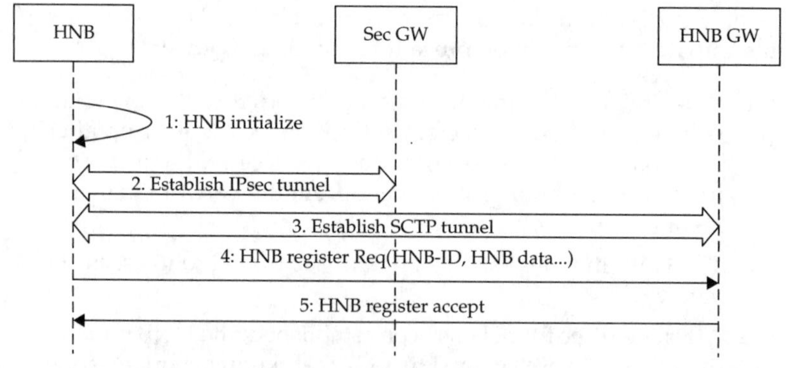

Figure 4-17 IuB Home NodeB registration

been accepted, the HNB gateway sends an accept reply back to the HNB. At this point the HNB is fully registered.

4.3.5 User Endpoint Registration

This section describes an IMS user endpoint registration scenario. After a HNB has successfully registered, it allows for one or more user endpoints to register with the HNB gateway. When a user endpoint either is turned on or moves into the proximity of a HNB, it attempts to register itself. Figure 4-18 shows a user endpoint registration procedure.

1. When a user endpoint moves into the proximity of an HNB, it initiates communication with the HNB by establishing a RRC connection with the HNB. Once the RRC channel is established, the UE sends its identity and its feature capability support along with additional information.
2. Once the RRC connection has been established, the user endpoint sends a RRC Initial Direct Transfer message requesting its location to be updated. This location update provides the mobile network the latest information on where the UE can be contacted; in this case, it is now via the HNB. Included in the Direct Transfer message is the User Endpoints Identification information.

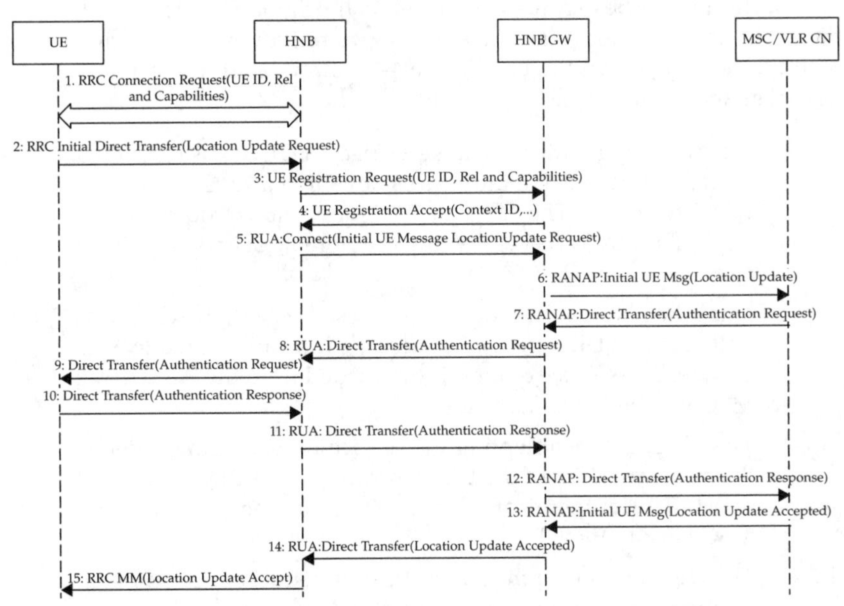

FIGURE 4-18 Iu-h UE registration

3. After the HNB obtains the UE's identity, it proceeds with sending a User Endpoint registration request to the HNB gateway. The registration request contains the UE's ID, its current release information, and its feature support capabilities.
4. The HNB gateway processes the registration request. If all the parameters are acceptable, including the feature support and version, the HNB gateway sends a registration accept response.
5. The HNB then sends a RUA connect message to the HNB gateway containing the Location Update Request.
6. After the HNB gateway processes the connect message, it sends a RANAP Initial Message containing the location update request to the MSC/VLR.

7–9. Before accepting the update, the MSC/VLR can request authentication of the user equipment by sending an authentication request to the UE.

10–12. The UE identifies itself by sending a Authentication Response message back to the MSC.

13–15. After the MSC validates the UE's identity it sends a location update accept message back to the UE. At this point, the UE has successfully registered with the network and can now process calls.

4.3.6 User Endpoint Call Origination over Iu-h

This section describes an Iu-h Femtocell call origination. After both the HNB and a UE have successfully registered, calls can be originated or terminated by the UE. Figure 4-19 shows a UE call origination over an Iu-h interface.

1–3. When a user attempts to make a call on a UE that is registered with an HNB, the UE initiates establishing a RRC connection with the HNB. The UE sends a connection request to the HNB. The HNB accepts the connection request and establishes a control channel.

4–6. Once the RRC connection has been established, the UE sends a RRC Initial Direct Transfer message indicating a request for service. The service request is forwarded from the HNB to the MSC via the HNB Gateway.

7–9. The MSC, before accepting the service request, can challenge the authentication of the UE by sending an authentication request. The authentication request is propagated from the MSC back to the UE.

10–12. The UE responds to the authentication request by sending an Authentication Response back to the MSC/VLR.

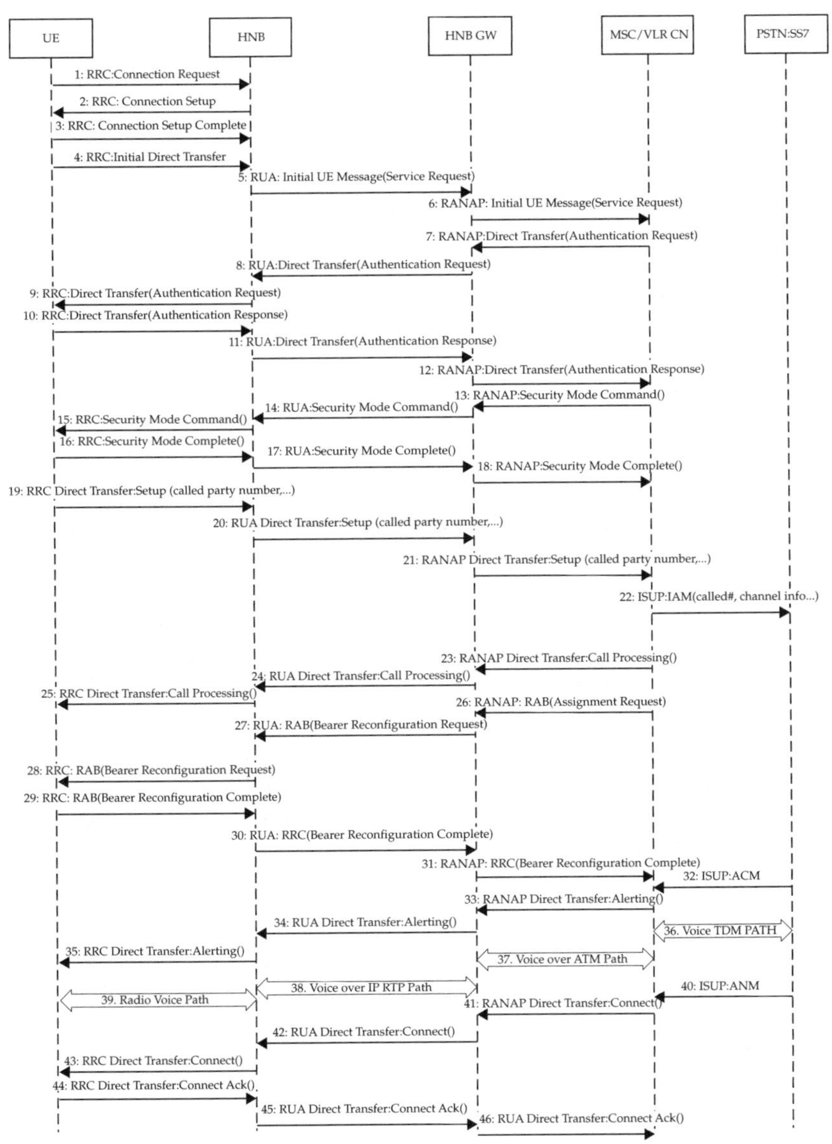

FIGURE 4-19 IuB UE call origination

13–15. After the MSC has validated the authentication, it can initiate ciphering encryption between the UE and the network. The ciphering prevents any third party from maliciously altering signaling messages, especially over the air interface. The MSC sends a RANAP Security Mode command to the UE to turn ciphering on.

16–18. When the UE receives the Security Mode Command, it responds to the MSC with a Security Mode Complete message indicating ciphering is turned on.

19–21. At this point the UE is ready to initiate a call with the network by sending a setup message using direct transfer signaling. The setup message sent from the UE to the MSC contains the called party number and other call-related parameters.

22. After the MSC processes and accepts the setup message, it determines the call is a PSTN call. The MSC sends a ISUP IAM SS7 message to establish a connection to the called party via the PSTN.

23–25. The MSC sends a call proceeding message back to the UE to inform it that the call is attempting to be set up toward the destination.

26–28. While the PSTN leg of the call is being established, the MSC initiates establishing a Radio Access Bearer (RAB) to carry the voice traffic over the air interface. The MSC sends a RAB assignment/reconfigure request to the UE.

29–31. Based on the information in the RAB request the HNB will allocate an appropriate radio channel. When the UE accepts the RAB assignment/reconfigure request, it sends a RAB complete reply back toward the MSC.

32–35. At some point the PSTN receives an indication that the called party phone is alerting (ringing), resulting in the MSC receiving an ISUP ACM (Alerting). The MSC notifies the UE that the far end is ringing by sending an alerting message.

36–39. The voice path is cut through on all segments from the PSTN to the radio interface of the UE.

40–43. At some point the called party answers the call, resulting in the MSC receiving an ISUP-ANM (answer). The MSC notifies the UE that the far end has answered by sending a Connect Message.

44–46. The UE accepts the connect message and sends a connect ACK back to the MSC. At this point the call is active and the two parties can have a dialog with each other.

4.4 LTE Femtocell Architecture

Long-Term Evolution (LTE) is an emerging wireless technology that many of today's service providers are planning to deploy. LTE is considered the next-generation technology in wireless communications, offering broadband access rates on wireless devices. This allows mobile wireless devices the ability to offer services that fixed-line broadband customers are accustomed to such as high-speed data, stereo-quality audio, and streamed high-definition TV (HDTV). Femtocell standards were first developed after 3G wireless standards had already matured. As a result, 3G femtocell standards issues were addressed after there was an opportunity to influence changes to the base standards for the benefit of 3G femtocell deployments. For LTE, femtocell standardization is part of the initial base standard [11] which allows LTE femtocell issues to be addressed early on in the development of LTE.

Femtocells may also assist in the deployment of LTE services from the perspective that the distribution of LTE NodeB radio transceivers will need to be significantly denser. Having to deploy significantly more NodeB radio transceivers spread across their coverage areas becomes a challenge, since there are a limited number of existing radio towers at their disposal. Using femtocells, service providers can extend coverage to any location where broadband access is available. This significantly increases a carrier's ability to provide LTE coverage in areas that were not feasible before.

Figure 4-20 shows a high-level diagram of a LTE network that includes a femtocell network. The mobile phone, also known as the user equipment (UE), interfaces with a Home eNodeB (HeNB) over-the-air interface. The Home eNodeB is a device that resides in the customer's home that is analogous to the Femto Forum's FAP. The

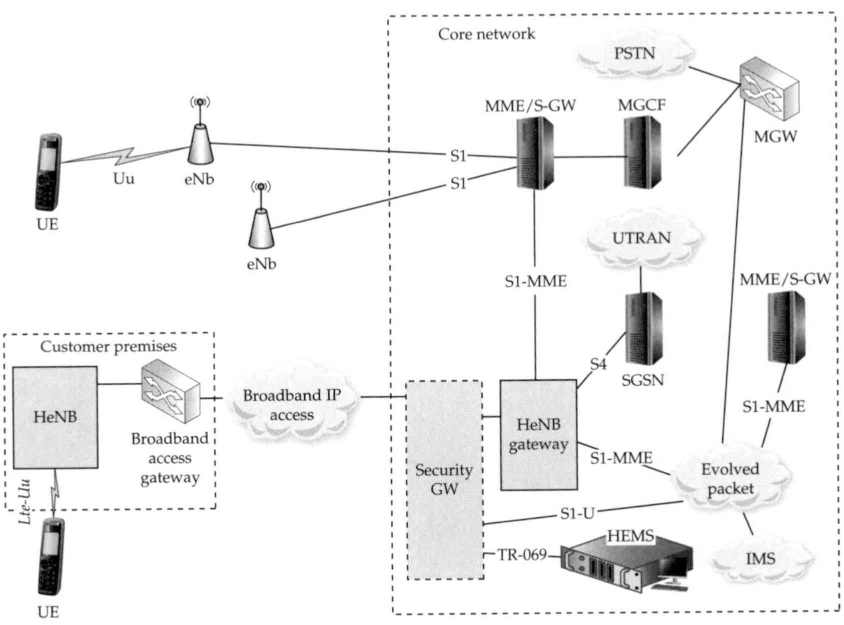

Figure 4-20 LTE Home NodeB network

HeNB is a miniature eNodeB providing wireless LTE coverage in the customer's home. The HeNB interfaces to the mobile network operator (MNO) over a broadband network by interfacing with a broadband access gateway.

The mobile operator's core network obtains access to the HeNB connected to a broadband access device via a security gateway. The security gateway is used to protect the core network against attacks. On the trusted side of the security gateway resides the Home eNodeB Gateway (HeNB GW). The HeNB GW is responsible for aggregating traffic from a large number of HeNBs and interfacing with the mobile operator's Evolved Packet Core (EPC) network. The S1-MME interface carries data using the LTE S1-MME interface. The HeNB interfaces with the EPC using a LTE S1-U interface via the security gateway.

LTE Femtocell Signaling Protocols

Figure 4-21 shows the LTE femtocell control protocol stack used by the LTE femtocell networking equipment.

The user equipment (UE) stack is the same stack the equipment uses to interface with enhanced NodeB, since the UE sees the HNB equivalent to an enhanced NodeB. At the physical layer is the Radio Frequency (RF) interface, also referred to as the air interface. Above the MAC layer the UE's higher-layer control protocols use the Radio Link Control (RLC) layer to gain access to various radio channels over which to communicate to the mobile network. The RLC provides error-free delivery of messages through mechanisms of error detection, message acknowledgments, message sequence number, and message retransmissions. The Packet Data Convergence Protocol (PDCP) performs header compression; increases packet delivery reliability; and provides ciphering, transfer of control plane data, and integrity protection.

The Radio Resource Control (RRC), as the name implies, is responsible for managing the radio resources that are allocated for

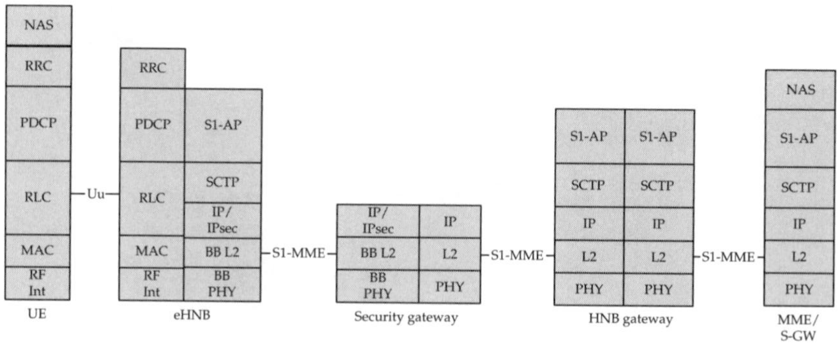

Figure 4-21 LTE femtocell control protocols

User Equipment. The RRC is responsible for establishing and releasing radio channels, including signaling and bearer channels. Radio signal measurements are invoked and reported by the RRC, allowing it to also perform mobility management. The mobility management includes determining if a call should be handed over, cell reselection, and routing updates. The RRC supports QoS, UE measurement reporting, and NAS message transfer. The NAS control protocol is responsible for EPS (Evolved Packet System) Bearer management, authentication, paging, and security control. EPS bearer channels over the radio interface are used for transporting media, including voice and data.

The S1 Application Protocol (S1-AP) is responsible for E-RAB management, LTE handover, RAT handover, paging, location reporting, load balancing, and overload control. The S1-AP protocol operates over the Stream Control Transmission Protocol (SCTP). SCTP, similar to TCP, is a transport-layer protocol providing reliable delivery of messages between the eHNB and the HNB gateway. At the IP layer, IPsec is used for security purposes, since the communications between the eHNG gateway and the security gateway is over an untrusted IP network. The Broadband Layer 2 (BB L2) and Physical Layer (BB Phy) depend upon what type of broadband access is being used by the subscriber, such as Fiber to the Home (FTTH), cable, or DSL. The HNB gateway is responsible for interfacing with the eHNB on the broadband access side via a security gateway. On other interfaces the HNG gateway interfaces with mobile service provider's core network over the S1-MME interface.

Figure 4-22 shows the LTE Femtocell Media protocol stack used by the LTE femtocell networking equipment. The user equipment (UE) stack is the same the stack it uses to interface with enhanced NodeB since the UE sees the HNB equivalent to an enhanced Node B. At the physical layer is the Radio Frequency (RF) interface, also referred to as the air interface.

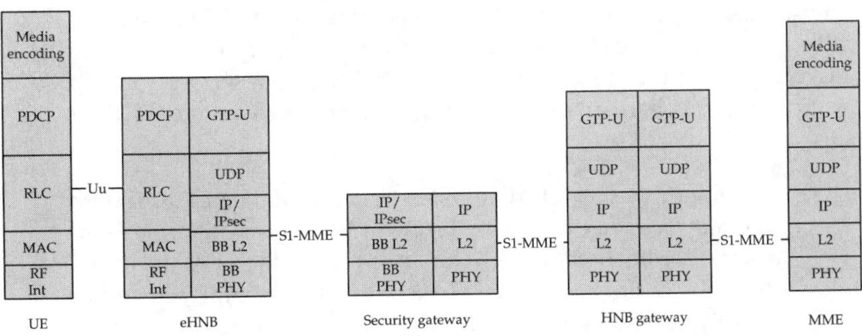

FIGURE 4-22 LTE femtocell media protocols

Above the MAC layer, the UE's higher-layer control protocols use the Radio Link Control (RLC) layer to gain access to various radio channels to communicate to the mobile network on. The RLC provides error-free delivery of messages through mechanisms of error detection, message acknowledgments, message sequence numbers, and message retransmissions. The Packet Data Convergence Protocol (PDCP) performs header compression; increases packet delivery reliability; and provides ciphering, transfer of control plane data, and integrity protection. At the IP layer, IPsec is used for security purposes, since the communications between the eHNB and the security gateway is over an untrusted IP network. The Broadband Layer 2 (BB L2) and the Physical Layer (BB Phy) depend upon what type of broadband access is being used by the subscriber, such as Fiber to the Home (FTTH), cable, or DSL. The HNB gateway is responsible for interfacing with the eHNB on the broadband access side via a security gateway. On other interfaces, the HNG gateway interfaces with the mobile service provider's core network over the S1-MME interface. The GTP-U is an IP-based protocol that allows for simple tunneling between nodes in the wireless network. Each tunnel allows for different QoS treatment, depending on the usage of the media.

4.5 Self-Organizing Femtocell Networks

Femtocells are deployed in large numbers scaling to the level of having a FAP in every home and business. As a result, FAPs will be installed by the end customer similar to the way customers install WiFi routers. When and where a customer installs a FAP is a completely random event. In order for femtocell operating costs to be kept to a minimum requires femtocell networks to be plug-and-play capable. To meet this need femtocells are designed to be SONs (Self Organizing Networks).

Macrocell radio networks are deployed with extensive radio measurements and parameter tuning manually performed by service provider employees. SON femtocells need to be able to adjust dynamically as they are installed and as conditions change resulting from neighboring femtocell installations. The following describes some of the possible mechanisms used that allow femtocell networks their SON capabilities:

- During power up a FAP registers and authenticates itself with the femtocell network. The FAP sends its location to the Femtocell Application Server. The FAS can send the FAP its initial tunable radio parameters based on its proximity to other FAPS and macro base stations.

- Once authenticated the FAP can automatically download the latest version of software, which includes fixes to known problems.
- The FAP should be capable of running self diagnostics to detect any hardware defects that would impact the performance of the femtocell.
- On regular intervals, the FAP can execute various radio measurements to detect changes in neighboring femtocells and macro base station transmitter power levels. After a power scan measurement is performed a FAP can adjust its tunable radio parameters to optimize its performance. These adjustments would be reported back to the FAS.
- The FAS can maintain an inventory of neighboring FAPs and macro base stations. Included in this inventory would be radio parameters such as transmit power and frequencies used. When changes in the network occur, such as a FAP being added, the FAS can request adjustments to one or more FAPs' radio parameters in order to optimize network performance in the proximity of change.

4.6 Femtozones

Femtocells create a new unique set of applications for the end customer. As a mobile subscriber enters the proximity of a femtocell, such as their home, they enter into a femtozone. A wireless device resides in a femtozone when the services being offered are via a femtocell network. There are significant customer benefits when in a femtozone, including the following:

- Lower tariffs for services since the back haul being used is the end customer's broadband service connection.
- Higher speed network access with significantly increased wireless bandwidth. This increase in bandwidth enables wireless devices' ability to offer services available on broadband networks, such as high definition video streaming. Applications can leverage the increase in bandwidth to exchange media between wireless devices and Application Servers that store media content, such as music and video.
- Presence information indicating when a subscriber enters a femtozone is available by the FAS. This presence information allows for new advance applications to be created. For example, family members can be notified whenever other family members enter or leave their home.

References

[1] Femto Forum, http://www.femtoforum.com.
[2] 3GPP, http://www.3gpp.org.
[3] 3GPP2, http://www.3gpp2.org.
[4] Broadband Forum, http://www.broadband-forum.org.
[5] 3GPP2 X.P0059-000-0 CDMA2000 Femtocell Network: Overview, Version 1.0, November 2009.
[6] 3GPP2 X.P0059-200-0 CDMA2000 Femtocell Network: 1X and IMS Network Aspects, Version 1.0, November 2009.
[7] 3GPP TS 25.467 UTRAN architecture for 3G Home Node B (HNB); Stage 2, June 2010.
[8] 3GPP TS 25.410 UTRAN Iu interface: General aspects and principles, Dec. 2009.
[9] 3GPP TS 25.468 UTRAN Iuh Interface RANAP User Adaption (RUA) signaling, April 2010.
[10] 3GPP TS 25.469 UTRAN Iuh interface Home Node B (HNB) Application Part (HNBAP) signaling, April 2010.
[11] 3GPP-TS 36.300 Evolved Universal Terrestrial Radio Access(E-UTRA) and Evolved Universal Terrestrial Radio Access Network (E-UTRAN); Overall description; Stage 2, April 2010.

CHAPTER 5

VoIP Fundamentals

In the last ten years the Voice over Internet Protocol (VoIP) has become one of the hottest technologies in telecommunications. This has occurred for many good reasons that will be discussed in this chapter. As VoIP has evolved, different signaling protocols have emerged, with one in particular (SIP) becoming the protocol of choice for the future and chosen for the 3G IMS architecture. These signaling protocols will be described along with a comparison of their fundamental differences.

5.1 Why VoIP

The Internet Protocol (IP) is the most widely used packet networking protocol to date by far. One of the first IP networks developed was the ARPANET, created by DARPA. DARPA (Defense Advanced Research Projects Agency) is an agency of the United States Department of Defense responsible for the development of new technologies. The early IP networks were used to connect computers together for exchanging data and information. By the early 1990s the majority of computers were networked using IP-based protocols of one form or another. Today the World Wide Web (WWW), which is built on IP, is used by people of all ages across the world for accessing web applications for all purposes of information exchange, including entertainment, banking, marketing, advertisement, transaction processing . . . (the list goes on).

While the use of IP networks was evolving, voice telephony was in a process of migrating its entire core infrastructure to a fully digital network. Included in the core infrastructure are digital switching systems and transmissions systems. Digital switching systems for conventional public telephone service providers include subscriber class 5 switches (5ESS, DMS100 . . .) and toll class 4 switches (4ESS . . .). For large and small business systems, digital PBXs and digital key systems became very popular. Digital transmission systems for telecommunication companies included T1 and DS3/T45 and fiber-optic transmission systems.

IP networks continued to grow at a rapid pace and were being used by most companies small and large. As a result, many IP products matured to allow for very reliable data networks. With its widespread usage, economies of scale drove prices of IP equipment down. IP networks were originally designed to carry data and were not capable of transporting real-time applications like voice and video. However, as the network performance improved, especially network delays, IP networks were able to carry real-time interactive applications such as voice and video far more cheaply and arguably more reliably than conventional TDM (Time Division Multiplexed) networks.

Commercial and residential access technologies during this period have also improved tremendously. The early days of Internet access for consumers were via dial-up modems that had data transfer rates of 19.2 and 9.6 Kbps. Dial-up modems are commonly referred to as narrow-band technology. Since then, various broadband access technologies have emerged, providing the consumer with multiples of 100 Kbps and mega-bps data rates. Some of the different broadband access technologies are

- DSL (Digital Subscriber Loop) is a broadband technology that operates over existing telephone wire known as twisted pair. This has the advantage of utilizing the huge amounts of copper wiring the local telephone companies have deployed over the last century.
- Cable Modem is a broadband technology that operates over the cable company's distribution plant, including the coax cable in the customer's home.
- FTTH (Fiber to the Home) uses various optical technologies to provide broadband capabilities to the customer over fiber-optic lines.
- WiMAX (Worldwide Interoperability for Microwave Access) uses wireless technologies to provide broadband service to the customer. WiMax's radio distances are typically in the range of one to five miles.

With the advancement of broadband technologies, consumers and businesses can connect to the public Internet at much higher data rates at reasonable costs. Companies large and small have a need for data networking of computers. These same companies have a need for voice networks. Within this last decade the IP networks have matured to support real-time applications like voice and video reliably with good quality. Since there are expenses to build, operate, and maintain both networks, having one network—namely IP—will reduce the cost by converging onto one infrastructure. This trend of migrating voice networks to IP is occurring at a steady rate and will continue as a result of the economics and technology advantages of IP networks.

This is why VoIP is and will continue to be a mainstay technology. This also holds true for other multimedia applications as well.

5.2 VoIP Signaling Protocols

With the advent of carrying voice over an IP network infrastructure, one necessity is to have a signaling protocol designed for VoIP. Signaling protocols are required to establish connections through a network in order to allow two peers to communicate over that connection. These are some important aspects of signaling protocols:

- Identification and authorization are needed to ensure that a device that is using the services of the network is truly a valid user. This is very important to prevent theft of service as well as preventing a customer's service usage from being hijacked.
- Establishing and releasing connections is the main purpose of a signaling protocol. As requests and responses traverse a network, resources are reserved and ultimately allocated along the way.
- Codec and application usage negotiation allows all parties involved in an application to agree on the various types of media they will exchange with each other. This negotiation is done on a call-by-call basis.
- Quality of service (QoS) provides a mechanism to give priority to real-time applications like voice and video in a packet-based network. QoS techniques minimize if not eliminate the human-noticeable degradation of service from having both real-time and non-real-time applications share the same network.
- Audits of resources are very useful to detect and proactively fix problems within the VoIP network.

5.2.1 VoIP Termination-Based Signaling Protocols

A conventional POTS (plain old telephone service) line is a copper-pair wire that connects on one end to the local telephone switch and the other to a subscriber's termination. A subscriber's termination typically is connected somewhere in the customer's home where the home electronic telephone is connected. Customers make POTS phone calls using electronic capabilities of a telephone device known as the BORSCHT functions. This acronym consists of the electrical interface requirements for a phone as follows: Battery feed, Overvoltage protection, Ringing, Supervision, Codec, Hybrid, and Testing.

The Supervision refers to supervisory signaling. There are several supervisory signaling types that are used, with loop start being the most common for subscriber lines. A loop start is a way a telephone signals the user's going off hook by closing an electronic switch allowing current to flow. When the user is not using the phone, the handset is on hook and the electronic switch is open, preventing current from flowing. When the user first goes off hook, the local telephone switch detects the current flow, then applies a dial tone, and then listens for DTMF (dual tone multifrequency) digits. Once the first digit is dialed, the local telephone switch removes the dial tone and waits to collect enough digits to be able to route a call. Once a call is routed to a destination, the called party is alerted by the local switch using a power ring. A power ring is an AC voltage signal at 20 Hz. The called user answers the call by going off-hook, and the two parties are connected via the local switch fabric cross-connect circuitry. Either party can terminate the call by going on hook.

VoIP termination-based protocols use IP messages that reflect the supervisor signaling used by POTS telephones. So for an example, when a user goes off hook, an IP message is sent from the VoIP phone to the network that states an off-hook event has occurred. The network then sends an IP message to the VoIP phone to apply a dial tone and collect digits. So the key attribute about VoIP termination protocols is that the messages mimic the electronic signaling that is found at the subscriber's termination.

The two most commonly used termination-based protocols are MGCP (Media Gateway Control Protocol) and H.248. Both will be described briefly in the sections that follow. Both protocols are very similar in their approach to solving the signaling requirements for VoIP. Megaco is the common protocol naming convention given to both MGCP and H.248. In general these Megaco control protocols are used between a media gateway (MG) and a media gateway controller (MGC), also referred to as a call agent. The protocol approach is a master-slave relationship whereby the MGC is in charge and the MG executes the commands the MGC instructs it to perform. MGCP is a result of the work of the IETF's Megaco working group, and H.248 is a result of the work of ITU-T Study Group 16. Although both protocols follow the same supervisory signaling approach, the protocols are different enough that they are not compatible with each other. The industry has also adopted different uses for each protocol that will be described in the relevant protocol sections to follow.

Although the MGCP and H.248 protocols are not compatible, they have enough similarities that protocol stack products usually bundle them into the same software code base. These are some of the key aspects that are similar between the two protocols:

- Both protocols use a string-based encoding. Both support a long form whereby most words are spelled out. They both support a short form where words are abbreviated.
- Both protocols use a transaction concept where a notification or command is sent using a transaction that needs to be acknowledged.

- Both have a media gateway controller or call agent that is the master and the media gateway that is the slave.
- A subscriber is identified with a termination ID that is unique to the media gateway.

MGCP

The MGCP protocol is specified in RFC 3435 [1], which replaces the original MGCP RFC 2705 [2]. In MGCP the call agent instructs the MG on what events should be reported, what signals should be applied to the headset or audible ringer, and what RTP stream the MG should send and receive voice packets on. A media gateway can be a small IAD (integrated access device) with one or two POTS lines or can also be a single-line phone. In some applications MGCP can be used by a LAG (large access gateway) whereby many POTS lines are supported.

MGCP supports two types of messages, where one is a command and the other is the response. Each command will contain a unique transaction ID. A response message will contain the same transaction ID so that the response can be correlated to the outstanding request. MGCP supports nine command verbs, each using a four-letter verb as follows:

- AUEP (Audit Endpoint) is a command the call agent sends to the MG to audit one or more endpoint terminations. This command is very useful during call agent startup to audit a MG for a list of all of its terminations. Note that there is one termination for every phone line supported by a MG. A call agent can also periodically audit individual terminations in order to validate their status.
- AUCX (Audit Connection) is a command the call agent sends to the MG to audit a specific connection. The call agent indicates what information about the connection it's interested in.
- CRCX (Create Connection) is a command the call agent sends to the MG to create a connection, which in turn prompts the MG to send and receive RTP packets. Some of the parameters of the connection are: the termination ID the connection is applied to, the call ID, and the SDP parameters of the connection.
- DLCX (Delete Connection) is a command the call agent sends to the MG to delete an existing connection. Included information in this message is the termination ID and the call ID the connection is associated with.
- MDCX (Modify Connection) is a command the call agent sends to the MG to modify an existing connection. Parameters that can be modified are: the send and receive modes, any aspect of the SDP, signals to be applied, and updates on which events the MGC wants to be notified of.

- RQNT (Request for Notification) is a command the call agent sends to the MG to requests events from the MG as they occur. Such events are the user going off hook, going on hook, or dialing a certain number of digits. In this command the call agent also has the ability to modify the signals being applied. This command also allows the call agent to define a sequence of events to collect and signals to apply.
- NTFY (Notify) is a message the MG sends to the call agent indicating a specific event has occurred. Examples of events the MG will notify the call agent of are: the user going off hook, the user going on hook, and digits dialed. These events are sent only if previously requested by the MGC.
- RSIP (Restart In Progress) is a message the MG sends to the call agent indicating that the MG is in the process of restarting. Contained in the message is the restart mode, which indicates the type of restart. This indicates if the MG is starting fresh and removing all termination connections or if it is in the process of going out of service.

MGCP Call Flow This section presents a MGCP basic call between two residential gateways with a single call agent coordinating the call. A residential gateway (RG), also known as an integrated access device (IAD), is a customer premises device located in the customer's home or office. An IAD supports VoIP on its IP interface and has one or more POTS lines. The residential gateways shown in Figure 5-1 use

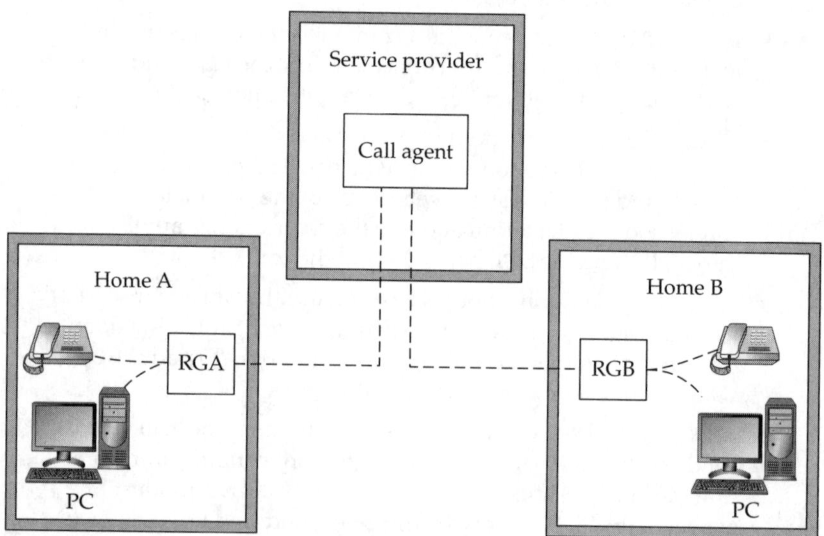

FIGURE 5-1 Residential gateway configuration

the MGCP protocol for establishing phone calls. The phone jack side supports the electrical BORSCHT functions described earlier so that standard telephone can be connected to the residential gateway, allowing the customer to place calls. Some residential gateways also provide broadband IP access for home computers, typically using an Ethernet interface. Other residential gateways require an access device like a cable modem that provides an IP interface.

The call flow shown in Figure 5-2 is a successful call establishment resulting in a conversation between customers A and B. The call agent could be a single networking device called a soft switch that typically resides in the central office. A soft switch is a generic computer server, without any customized telephony hardware, that supports functions of signaling, routing, and billing. Figure 5-2 shows a basic MGCP call flow, and a description of each message is provided. This basic call shows a call placed from Home A to Home B. All of the MGCP protocol messages exchanged between the residential gateways and the call agent are shown.

The following describes each of the messages that are exchanged in numbered order:

1. Residential gateway (RG) A sends a Restart In Progress (RSIP) to the call agent to indicate that the RG A is starting up.
2. The call agent processes the RSIP by clearing any outstanding activities with RG A, and sends an ACK to RG A.
3. The call agent now sends a wild-card audit endpoint message to RG A. The wild-card audit requests RG A to provide a list of all endpoints/terminations that RG A has provisioned. Note that a residential gateway can support one or many terminations.
4. RGA acknowledges the audit endpoint and provides the list of terminations it has provisioned.
5. The call agent sends a request for notification informing RGA that it should send an event as soon as the user goes off hook.
6. RGA sends an acknowledgment indicating that it is armed and waiting for a user to go off-hook.
7. Same as Step 1 for RGB.
8. Same as Step 2 for RGB.
9. Same as Step 3 for RGB.
10. Same as Step 4 for RGB.
11. Same as Step 5 for RGB.
12. Same as Step 6 for RGB.
13. At some point user A picks up the handset on the phone to place a call. This causes RGA to send a notify message indicating that an off-hook event has occurred.

14. The call agent acknowledges it received the off-hook event.
15. The call agent instructs RGA to inform it of any collected digits or on-hook events. RGA is also instructed to apply dial tone signal so that the user hears dial tone.
16. The call agent acknowledges the receipt of the notification event.
17. After the user enters the telephone number for B, RGA will send a notify message containing the digits that were dialed. Note RGA knows the number of digits to collect based on a digit map of information that is provided by the call agent.
18. The call agent acknowledges the receipt of the digits dialed.
19. The call agent sends a RQNT message to RGA requesting to be notified if the user goes on hook.
20. RGA accepts the RQNT and sends an ACK to the call agent.
21. The call agent sends a CRCX message to create a local connection at RGA. Included in this message are the connection requirements like the codec type to be used and the packetization period.
22. RGA accepts the request to create a connection by acknowledging the CRCX. The ACK includes the connection information chosen by RGA, including the IP address and port the voice packets will be sent from.
23. The call agent sends a CRCX message to create a connection at RGB. Included in this message is the RGA connection information provided in the message 22.
24. RGB accepts the request to create a connection by acknowledging the CRCX. The ACK includes the connection information chosen by RGB, including the IP address and port the voice packets will be sent from.
25. The call agent sends a MDCX message to RGA to modify the connection created in message 21. The MDCX will set the remote SDP for RGA that was provided by RGB in message 24. This prepares RGA to have all information necessary about RGB so that it can prepare its RTP stream for voice traffic.
26. RGA accepts the MDCX message and sends an ACK message in response.
27. The call agent sends the RGA a RQNT request to send it a notification if user A goes on hook. In the same message it instructs RGA to play a ring tone letting user A know that user B's phone is ringing.
28. RGA accepts the RQNT message and sends an ACK message in response.

29. The call agent sends the RGB a RQNT requesting it to send a notify if user B goes off hook in order to start the call. In the same message it instructs RGB to play a ring tone letting user A know the user B's phone is ringing.
30. RGB accepts the RQNT message and sends an ACK message in response.
31. User B eventually decides to answer the phone and lifts up the handset. RGB then sends a notify message indicating an off-hook event has occurred.
32. The call agent accepts the notify and replies with an ACK.
33. The call agent sends the RGB a RQNT requesting it to send a notify if user B goes on hook to end the call or if user B presses flash-hook to put the current call on hold.
34. RGB accepts the RQNT message and sends an ACK message in response.
35. The call agent sends the RGA a RQNT requesting it to send a notification if user A goes on hook to end the call or if user B presses flash-hook to put the current call on hold.
36. RGA accepts the RQNT message and sends an ACK message in response.
37. The call agent sends the RGA a MDCX message to modify its connection mode to send and receive. This allows RGA to send and receive packetized speech on the agreed-upon RTP connection.
38. RGA accepts the MDCX message and sends an ACK message in response.
39. A two-way communication VoIP channel is now established, and a voice conversation can now occur between user A and user B.

H.248

H.248 is very similar to MGCP protocol in the messaging and the approach to establishing connections. Both protocols use reporting of events that occur at the physical POTS interface; however, the H.248 protocol enhances the control of the MGC to allow for more flexibility and extensibility. H.248 defines a *physical* termination that uniquely identifies an electrical interface on the gateway, which can include a POTS interface or a channel on a DS1 interface. An *ephemeral* termination is a temporary connection established for sending and receiving a RTP stream for the duration of the call. A call is created by forming a context that connects physical and ephemeral terminations.

Similar to MGCP, H.248 uses a transaction ID when sending a message in order to correlate the response message with the request.

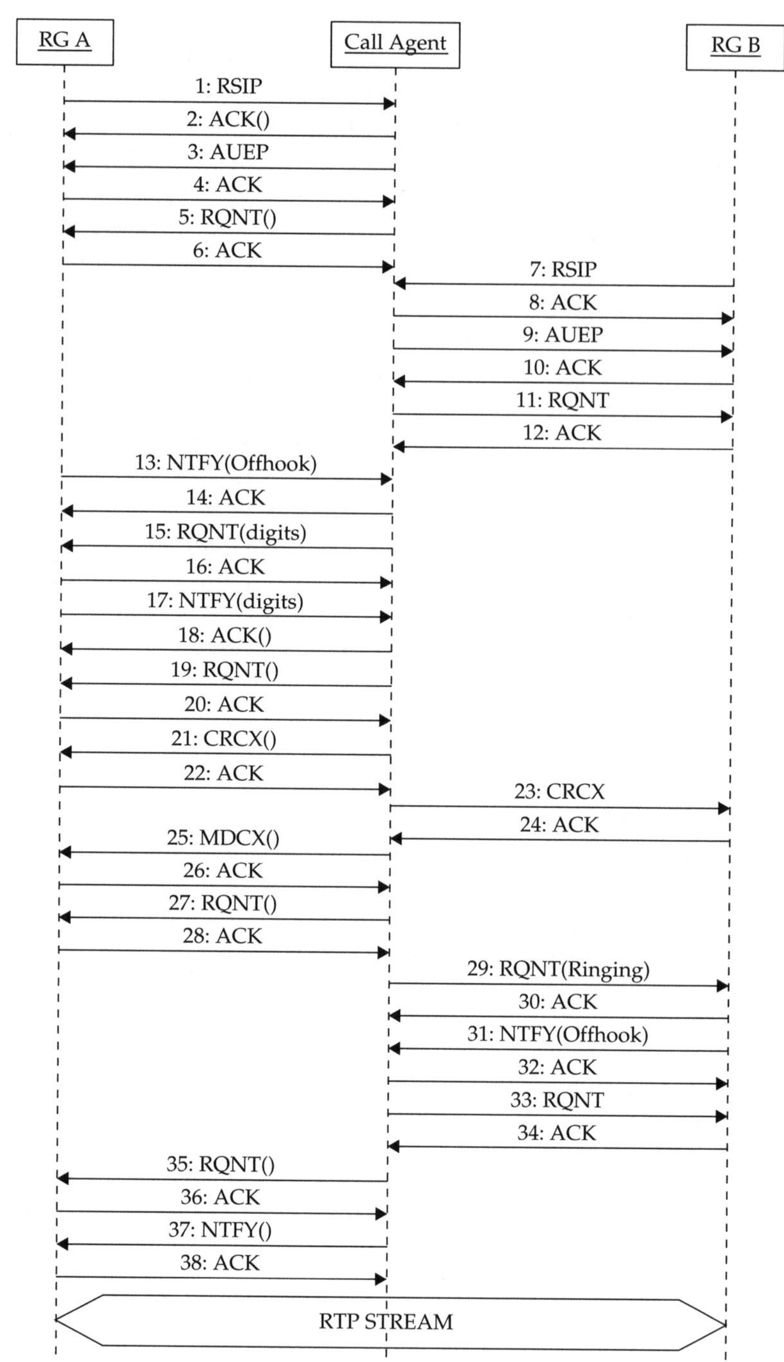

FIGURE 5-2 MGCP basic call flow

H.248 uses a nested message structure whereby one message can contain multiple transactions. A transaction then can contain multiple commands about a context. A command is further composed of descriptors that allow for extensible detailed information about commands and their responses. This added message complexity allows H.248 to support LAGs and multimedia connections, including voice, video, and data conferencing.

The following provides a summary of the command types supported by H.248:

- **Add** The add command adds a termination to a context. The termination can be physical or an ephemeral.
- **Modify** The modify command modifies an existing context. The parameters that can be modified are the same as the parameters for the add command.
- **Subtract** The subtract command removes a termination from a context. The termination being subtracted can be physical or ephemeral.
- **Move** The move command moves a termination from one context to another in one step. This command is useful for call waiting scenarios when a user toggles between two or more calls.
- **Audit Value** The audit value command retrieves the current values of properties, events, signals, and statistics associated with terminations. This command is useful in validating the state of terminations between the MGC and the MG.
- **AuditCapabilities** This command allows the MGC to query the MG on all its possible values it supports for signals, events, properties, and statistics for the different terminations.
- **Notify** This command allows the media gateway to inform the media gateway controller of the occurrence of events in the media gateway. Some common events are on hook, off hook, flash hook, and dialed digits.
- **ServiceChange** This command is used by both the media gateway and the media gateway controller to inform the peer of a change in operational status of one or more terminations. The media gateway can inform the media gateway controller of terminations going either into or out of service using the service change command. The media gateway also uses the service change command to initially register with the media gateway controller when it first powers up. The media gateway controller uses the service change command to instruct the media gateway to place terminations in or out of service. The media gateway controller can also use the service change command to inform the media gateway that a new MGC is taking over control.

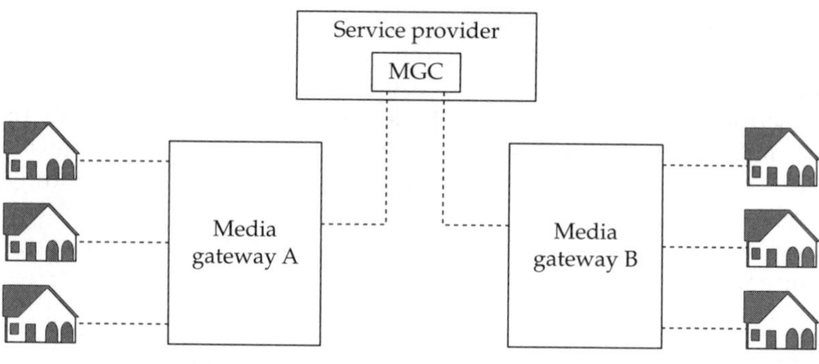

Figure 5-3 Media gateway configuration

H.248 Call Flow H.248 is usually used by devices that service multiple customer terminations. Such devices are called media gateways or MSANs (multiservice access nodes). A VoIP MSAN can use H.248 on its IP access side and have multiple POTS, ISDN, and DSL interfaces on its subscriber side. Figure 5-3 shows media gateways that reside outside the home, are local to a residential area, and service multiple customers. This is different than the MGCP example shown in Figure 5-1, where the residential gateway provided service to only a single customer.

The call flow shown in Figure 5-4 is a successful call establishment resulting in a conversation between customers A and B. The MGC (media gateway controller) could be a single networking device that is called a soft switch that typically resides in the central office. A *soft switch* is a generic computer server, without any customized telephony hardware that supports functions such as signaling, routing, and billing. Figure 5-4 shows a Basic H.248 call flow, and a description of each message is provided. This Basic Call shows a call placed from Home A to Home B. All of the H.248 protocol messages exchanged between the media gateways and the gateway controller are shown.

The following describes each of the messages that are exchanged in numbered order:

1. In this scenario the media gateway controller (MGC) has rebooted and sends a Service Change Message restart message to the Media Gateway A (MGA). If this scenario was media gateway rebooting, MGA would send a Service Change restart to the MGC.
2. The MGA accepts the service change and sends an acknowledgment message.
3. The MGC sends a modify message to MGA requesting that it send a notify message if the customer goes off hook. It also sets the physical termination mode to send and receive in preparation for a phone call.

4. The MGA processes and accepts the modify from MGC by sending a reply acknowledgment message.
5. Similar to Step 1 the MGC sends a Service Change Message restart message to the Media Gateway B (MGB).
6. The MGB accepts the service change and sends an acknowledgment message.
7. The MGC sends a modify message to the MGB requesting that it send a notify message if the customer goes off hook. It also sets the physical termination mode to send and receive in preparation for a phone call.
8. The MGB processes and accepts the Modify message from the MGC by sending a reply acknowledgment message.
9. At some point a person in a home served by the MGA decides to make a phone call and lifts the handset on his house phone. This event is detected by the MGA, which then sends a Notify message indicating an off-hook event has occurred. Included in that message is the termination identifier that corresponds to the customer that went off hook.
10. The MGC accepts the Notify message and sends an acknowledgment reply message.
11. The MGC sends a Modify message to MGA to apply a dial tone. The Modify message includes instructions to collect digits (phone number) using the dial plan rules provided by a digit map.
12. The MGA accepts the Modify message and sends an acknowledgment reply message.
13. The person placing the phone call enters a phone number after hearing a dial tone, which results in the MGA sending a Notify message containing the digits dialed.
14. The MGC accepts the Notify message and sends an acknowledgment reply message.
15. The MGC sends the MGA an Add message to add an emphemeral connection. Also specified in the Add message is the SDP template to use to pick SDP values for.
16. MGA accepts the Modify message and sends an ACK reply containing the local SDP that it chose.
17. The MGC obtains a route for the entered phone number and sends an Add to MGC B. The Add includes the termination ID for the destination phone and the SDP provided by MGA in Step 16. The Add message also instructs the MGB to choose a local SDP and to ring the phone for the termination destination termination.

18. MGB accepts the Add message and sends an ACK reply containing the local SDP that it chose.
19. MGC sends a Modify message to the MGA providing the SDP it received from the MGB in Step 18.
20. MGA accepts the Modify message and sends an ACK reply.
21. At some point the person being called hears her phone ringing and decides to answer the phone by picking up the handset. The MGB detects this event and sends a Notify message indicating the off-hook event. Included in the message is the unique termination identifier for the customer that just answered the phone.
22. The MGC accepts the Notify message and sends an acknowledgment reply message.
23. The MGC sends a Modify message to MGB requesting that it be informed if the user goes on hook. The Modify message also sets the mode to send and receive with no tones applied, preparing for a two-way conversation.
24. MGB accepts the Modify message and sends an ACK reply.
25. The MGC sends a Modify message to MGA setting the mode to send and receive with no tones applied preparing for a two-way conversation.
26. MGB accepts the Modify message and sends an ACK reply. At this point both calling and called parties have a two-way communication RTP channel established.
27. A conversation takes place between the calling and called party users.
28. At some point the called party decides to end the conversation by hanging up. MGB detects this event and sends a Notify message indicating the off-hook event. Included in the message is the unique termination identifier for the called party.
29. The MGC accepts the Notify message and sends and an acknowledgment reply message.
30. The MGC initiates tearing down the call by sending a Subtract message to MGB.
31. MGB accepts the Subtract message and sends an ACK reply.
32. The MGC sends a Subtract message to MGA to tear down the other leg of the call.
33. MGB accepts the Subtract message and sends an ACK reply. At this point the phone call has been fully torn down.

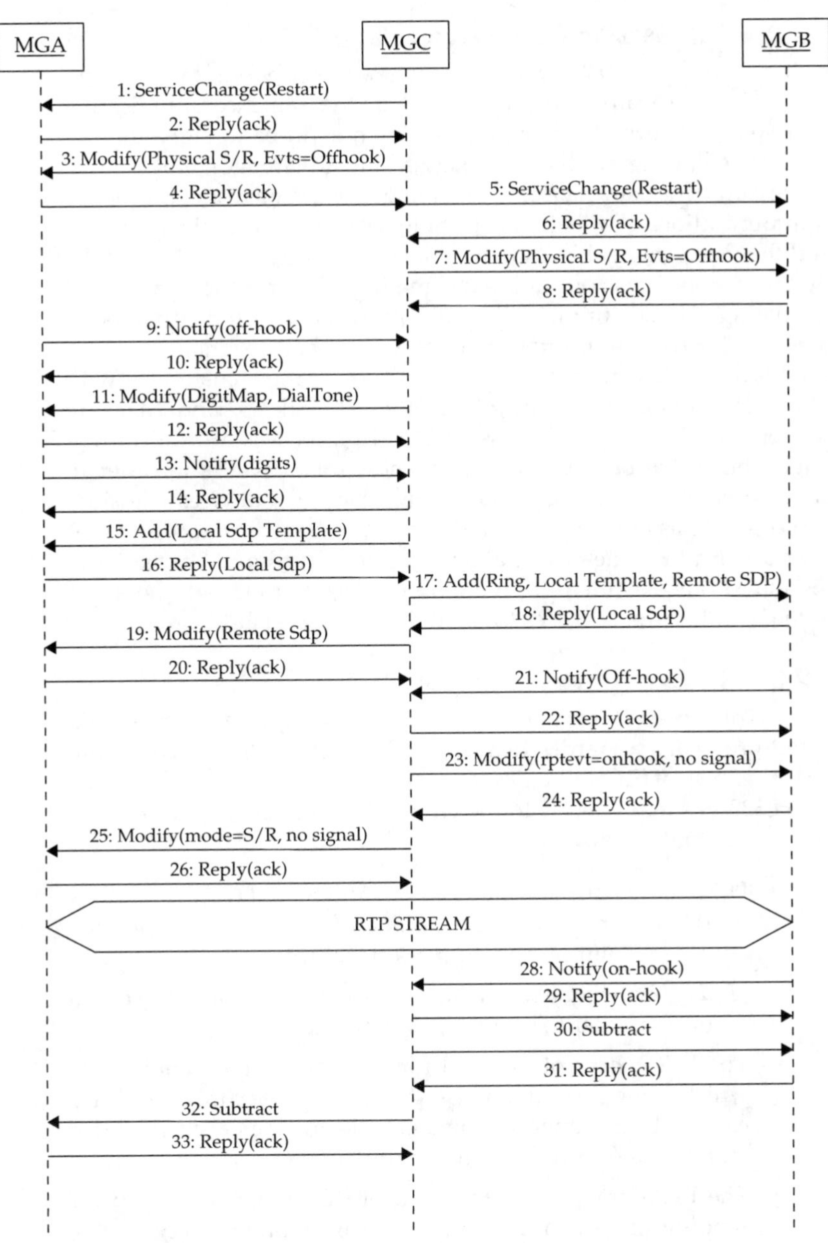

FIGURE 5-4 H.248 Basic Call Flow

5.2.2 VoIP Session-Based Protocols

In the early years of telephony networks the supervisory signaling and the voice traffic were contained in the same communication channel. This form of signaling is known as in-band signaling. As modern computers advanced, significant advantages emerged to processing the supervisory signaling separate from the voice communications channel. This technique of separating the signaling and the voice channel is known as out-of-band signaling. One of the first protocols to use out-of-band signaling was a protocol called SS7 (Signaling System number 7). This protocol is still in use today between telecommunication carriers. Out-of-band signaling is also used today between service providers and business customers with ISDN (Integrated Services Digital Network). Both SS7 and ISDN are session-based protocols. A session-based protocol is a protocol that establishes a temporary connection known as a dialog in order to have interactive communications between two devices. Session-based protocols are distinct from termination-based protocols in that they don't mimic the electrical supervisory signaling. Alternatively, session protocols attempt to minimize the number of messages exchanged between devices for better network scalability.

5.2.3 H.323

When Voice over IP became a realistic technology, the first session-based protocol standardized was H.323, which is also derived from the principles of SS7 and ISDN.

H.323 has many facets to the entire protocol suite, including but not limited to the following:

- H.225.0 Registration, Admission, and Status (RAS), which is used between an H.323 endpoint and a gatekeeper to provide address resolution and admission control services.
- H.225.0 Call Signaling, which is used between any two H.323 entities in order to establish communication.
- The H.245 control protocol for multimedia communication, which describes the messages and procedures used for capability exchange; opening and closing logical channels for audio, video, and data; and control and indications.
- The Real-Time Transport Protocol (RTP), which is used for sending or receiving multimedia information (voice, video, or text) between any two entities. RTP is used by most, if not all, VoIP protocols to carry voice conversations over an IP network.

H.323 is a message-based protocol that is binary encoded using Type, Length, and Value (TLV) fields. The following is a brief summary of some of the signaling messages:

- **Setup and Setup Acknowledge** The setup message is used to send all of the desired attributes of the connection being requested. Included are the calling and called parties, the connection bandwidth requirements, the QOS parameters of the connection, and a description of the RTP to be used.
- **Call Proceeding** Call proceeding informs the originator that the call request is being processed.
- **Connect** This indicates that the call has been answered.
- **Alerting** This indicates that the called party is being altered (phone is ringing).
- **Release Complete** The release complete message indicates that the call and its corresponding connection have been cleared.
- **Status and Status Inquiry** The status inquiry is used to query the state of the call from the peer entity. The status is the message used to provide the call state information.

5.2.4 SIP Introduced

SIP stands for Session Initiation Protocol, which has become the VoIP signaling protocol of choice. It is the signaling protocol chosen for the 3G IMS architecture, which will be described in a later chapter. SIP was originally modeled after the very successful HTTP protocol. HTTP has proven itself to be a very scalable and reliable protocol operating on IP networks.

SIP allows establishing and deleting multimedia sessions on an IP network. Some of the types of multimedia sessions include streaming of voice, video, and fax. SIP can support applications that include voice calls, video calls, voice and video conferencing, presence event notifications, instant messaging, and gaming sessions, to name just a few. Calls can be the default two-party or multiparty call sessions that can include one or several media streams.

SIP was originally defined by IETF, and its complete specification can be found in RFC 3261 [3]. SIP is very feature rich, with many RFCs in addition to 3261 that enhance the base protocol for functionality such as notifications and relocation capabilities. SIP is also the signaling protocol used in the 3GPP IMS architecture for multimedia IP connections supported by cellular networks. This is why SIP is chosen as a key component of femtocell technology.

5.2.5 SIP Details

SIP uses a peer-to-peer communication model whereby the endpoints are intelligent devices. This is a significant difference compared to MGCP and H.248, where the endpoints strictly follow the instructions of the media gateway controller (MGC). Using a peer-to-peer model enables endpoints to implement more advanced features and minimizes the resources needed by the network. SIP supports all of the features one expects from a public switched voice network and allows for more advanced multimedia features and applications, maximizing the benefits of an IP network.

Many session-based protocols have been defined by ITU, including H.323 (VoIP), SS7, ISDN (Q.931), and ATM (Q.2931). SIP, however, is a session-based protocol proposed and standardized by the IP community within the IETF. The SIP protocol can run on just about any transport-layer protocol. Since it typically runs on IP and has the ability to retransmit unacknowledged messages, it usually runs on UDP. Using UDP has the least overhead and allows for simple implementations of link redundancy strategies. Other popular transport protocols SIP runs on are TCP and SCTP. Figure 5-5 shows the relationship of SIP and the transport protocol, options being SCTP, TCP and UDP. Also shown is a codec's packetized voice being transported using RTP. All of the mentioned protocols are carried over the IP protocol.

SIP uses a text-based encoding, similar to MGCP and H.248, allowing humans to directly read the protocol PDUs. SIP, like other IP-based VoIP protocols, uses SDP (Session Description Protocol) to signal the desired multimedia connection. Typically the connection that is established uses RTP (Real-Time Transfer Protocol) to stream the media content (voice, video, fax . . .). SIP was first defined in the IETF RFC 2543 [4]. The protocol was enhanced using RFC 3261, and numerous other SIP-related RFCs exist, extending SIP for the purpose of offering additional features.

SIP Messages

There are two types of SIP messages, one a request and the other a response. Both types of messages contain a start line, followed by one or more header fields, followed by a CRLF (carriage return, line feed),

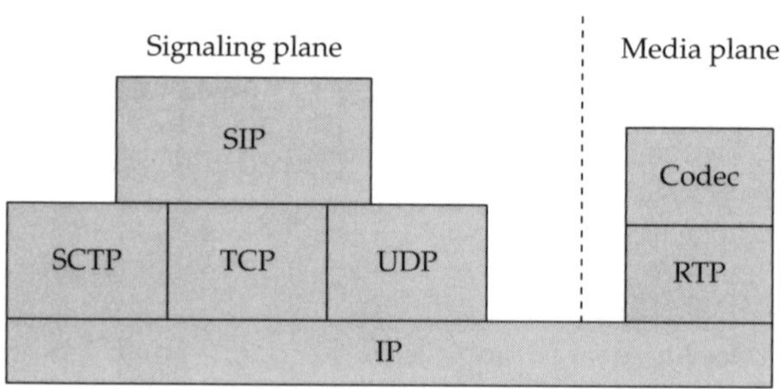

Figure 5-5 VoIP protocol stack

followed by an optional message body. The following is a BNF (Backus Normal Form) representation of a message:

start-line *message-header CRLF [message-body]

where CRLF = carriage return line feed.

A SIP request message will use a request line for the start line. The request line contains a request method that will be described. A SIP response will use a status line for the start line. The status line provides a summary of the processing of the request. A header field provides information about the message in the form of a name-value pair format. The name comes first, followed by a colon separator and then the value. It is possible to have multiple values for a field name using a semicolon to separate the multiple values. Also header name/value pairs are allowed to occur on multiple lines. For example, Figure 5-6 shows three equivalent header name/value pairs for Remote Party ID.

SIP Requests A SIP Request uses a request line for the start line described in the preceding section. The request line looks like the following:

Request-Line = Method SP Request-URI SP SIP-Version CRLF

where SP = space and CRLF= carriage return line feed.

The method is essentially a desired operation to be performed. The base SIP RFC 3261 defines the following six methods:

- REGISTER is a method used for registering contact information in the network such that the network knows what address and device to send requests to based on the Public Address of Record. The Public Address of Record is the address at which a person is known to be reachable. Today this is usually a telephone number. In the future this may be a name identity similar to an e-mail address.

Remote-Party-ID: <sip:phoneA@example.com>
Remote-Party-ID: <sip:phoneB@example.com>
Remote-Party-ID: <sip:phoneC@example.com>

3 Remote Party IDs on 3 Lines

Remote-Party-ID: <sip:phoneA@example.com>, <sip:phoneB@example.com>
Remote-Party-ID: <sip:phoneC@example.com>

3 Remote Party IDs on 2 Lines

Remote-Party-ID: <sip:phoneA@example.com>, <sip:phoneB@example.com>, <sip:phoneC@example.com >;id-context=plain

3 Remote Party IDs on 1 Line

Figure 5-6 Remote party ID representations

- INVITE is a method used to initiate or refresh/update a call session. The INVITE contains all of the specifics about the desired call to be established, including details of the originator and the destination being called.
- ACK is a method that acknowledges an INVITE request and is typically the final response that confirms that the media will start to stream end to end.
- CANCEL is a method used to cancel any outstanding requests. If a device detects or decides to no longer attempt to establish a dialog with the destination, it will send a cancel. An example scenario is a user dialing a number and then immediately hanging up because she changed her mind.
- BYE is a method used to terminate a call/session and can be sent by any (calling or called) party. This usually occurs when either party hangs up during the life of a conversation.
- An OPTION is a method used to query the peer of its supported messages. This helps a SIP device (endpoint or switch) to know if its peer supports different capabilities. The OPTION method is also used for a keep-alive message between two devices. This is useful to know the status of the communications between two devices at the SIP protocol level.

Several RFCs have defined additional SIP methods for the purposes of extending the capabilities of the SIP protocol. Some of these RFCs are

- RFC 3262 [5] (Provisional Response Acknowledgment Method) PRACK extends SIP to allow for responses to be acknowledged. This was added because RFC 3261 by default didn't allow for responses to be acknowledged, which caused problems in some scenarios. For example a 180 Ringing response to an INVITE provides the status that the far end is ringing. If this response is not received by the originator, a ringback tone may not be played to the calling party. The PRACK method allows for a response such as 180 to be acknowledged, ensuring the proper interaction with the user.
- RFC 3265 [6] (SIP Subscription and Events) extends SIP to allow a SIP device to subscribe to a notification service. This is very useful when events occur in a network and a device can present an indication to the end user. One popular usage is a SIP phone subscribing for voicemail notifications. When a voicemail is left for the subscriber, a notification is sent to the SIP phone, allowing it to indicate a voicemail message exists by lighting a lamp on the phone, or even playing an audible tone. Subscriptions can also be used to request information about status of other subscribers, which is very useful for business systems that support sharing of line appearance status.

- RFC 3903 [7] (SIP Extension for Event State Publications) added a Publish mechanism allowing a device to publish data that another device can subscribe to and get the published data via notifications.
- RFC 2976 [8] (SIP INFO Method) extended SIP to have a message that carries additional application information end to end. This method does not change the state of a session in terms of the existing media stream. The INFO method is useful for mid-call information exchanged such as additional digits collected. The mid-session information can be communicated either in an INFO message header or as part of a message body.
- RFC 3515 [9] (SIP Refer Method) adds the refer method that allows an active call to essentially be transferred to another party. When a replaces header is included in the Refer message, then the call is transferred to an existing call. This can be used for a consultative transfer where the party being transferred to is called first and then verbally asked permission to accept the call. If a replaces header doesn't exist in the Refer header, then the call is transferred by the network, establishing an add-on call prior to transfer. This can be used to support blind transfer where an active call is transferred without asking for verbal acceptance of the call.
- RFC 3311 [10] (SIP Update Method) is used to update the parameters of a call in progress that has not yet become active. For a stable active call a Re-INVITE method would be used to modify a connection. Both methods are commonly used to update SDP as part of the offer-answer model described later in the chapter.

SIP Responses SIP responses are sent as a result of processing a request. SIP response codes are consistent with response codes used in HTTP sessions. Based on the responses received, the call will either progress successfully to an active call or be released. The following summarizes the categories of responses that can be received:

- A 1xx is an Informational Response whereby the sender of the response is continuing to process the request. This response indicates that the server contacted is performing some further action and does not yet have a definitive response. A server sends a 1xx response if it expects to take more than 200 ms to obtain a final response. Also this response is not acknowledged. This is when a PRACK is used if it is required to guarantee the delivery of this response. Some examples of 1xx are: 180 means the far end phone is ringing; 181 means the call is being forwarded; 183 means the call is progressing.

- A 2xx response indicates that the request was successfully received and accepted. For example, 200 OK means the request was accepted; if the request was an INVITE, the 200 OK means the call was answered and a conversation can begin.
- A 3xx response means that further action needs to be taken in order to complete the request. This response is usually used during call establishment to redirect the call setup to another destination. This can be used for forwarding the call for the purpose of load balancing or as an overload redirection mechanism.
- A 4xx response indicates that the request failed, with the specific 4XX number identifying the reason. The following are some examples: 400 means Bad Request (the request received had something wrong with it), 404 – User Not Found (the phone number dialed is not a routable number), 480 – Temporarily Unavailable (the called device is not ready to accept calls), 483 – Too Many Hops (the call has traversed too many network nodes).
- A 5xx response indicates the request was valid; however, the receiver equipment could not honor the request and the exact number provides the best description of the problem. The following are some examples: 500 – Internal Server Error (the called equipment is having problems answering calls), 501 – Not Implemented (the SIP request method is one the device doesn't support), 502 – Bad Gateway, 503 – Service Unavailable.
- A 6xx response indicates that there is a Global Failure. Some examples of this error response are: 600 – Busy Everywhere (this device will fail all calls, since it's busy), 603 – Decline, 604 – Does Not Exist Anywhere, 606 – Not Acceptable.

SIP Call Flow

This section shows a SIP basic call between two users using a SIP telephone. Figure 5-7 shows a simple SIP proxy being used by the service provider that performs the function of a call server. The SIP proxy could be a networking device that is called a soft switch that resides in the central office. A soft switch is a generic computer server, without requiring any customized telephony hardware that supports functions such as signaling, routing, and billing. User A and user B would reside in the customer's home office with an access router connected to a SIP phone and a home computer.

Figure 5-8 provides a simple SIP call flow that shows the SIP signaling for a call being established, followed by a conversation and then the call being torn down. All of the SIP protocol messages exchanged between the SIP phone and the SIP proxy are shown, and a description of each message is provided.

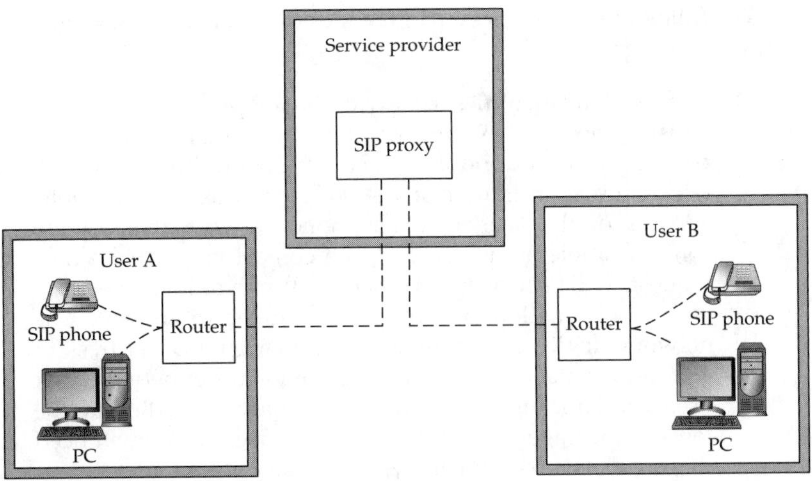

FIGURE 5-7 SIP service provider configurations

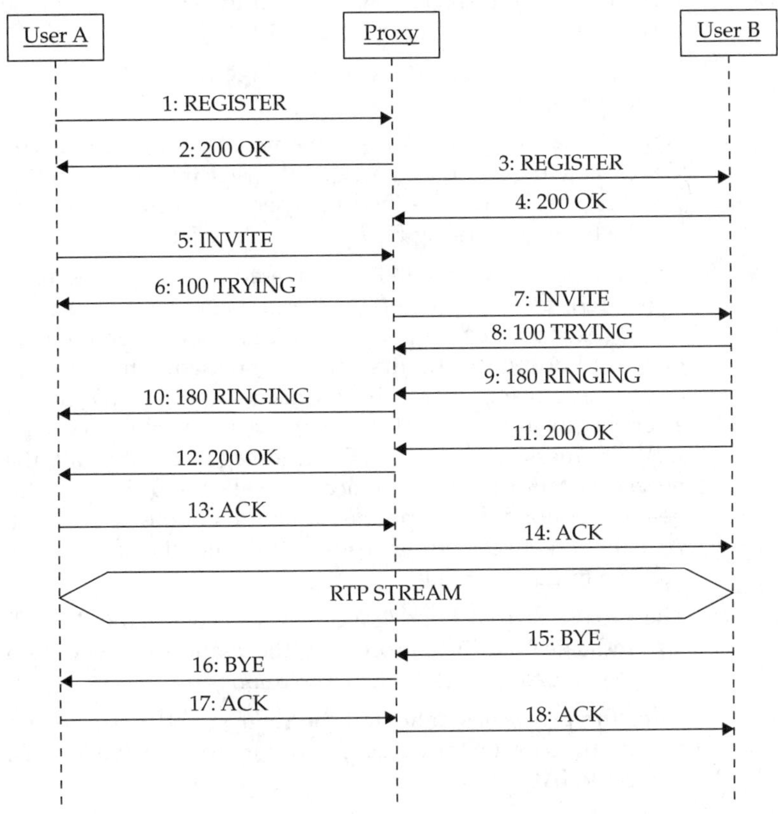

FIGURE 5-8 SIP basic call

The following describes each of the messages that are exchanged in number order:

1. User A's IAD identifies itself with the SIP proxy by sending a register message. A register message contains the user's address of record and a contact list. The address of record, which comes in the register's "From" header, is the public address of the user. In a telephone network this usually contains a telephone number. The contact list covers where requests will be forwarded when a call arrives in the network with a destination matching the address of record. The network device that maintains the mapping of address of record to contact list is referred as a registrar. A network that uses authentication will have additional steps whereby the register message is challenged and the user A IAD will need to provide a successful encrypted password.
2. When the proxy accepts the registration, it sends 200 OK to state the fact that it accepts the registration. From this point forward, calls destined for user A's address of record will be forwarded using A's stored contact list.
3. This is the same as register from message 1, except it has user B's information.
4. Again when the proxy accepts the registration, it sends 200 OK to state the fact that it accepts the registration. From this point forward, calls destined for user B's address of record will be forwarded using B's stored contact list.
5. At some point after the IAD is powered up and successfully registered, the user goes off hook and dials a phone number. A SIP IAD/phone will play a local dial tone when the user goes off hook until the first digit is pressed. Once enough digits have been entered by the user, the SIP IAD/phone attempts to initiate a call by sending the proxy server an INVITE message. The INVITE message contains all the necessary information to place the call, including the "to header" containing the phone number trying to be reached. Also included is the "from header," the caller ID information, and the initial offer SDP.
6. The proxy server acknowledges that it received the INVITE and indicates that it is processing the request by sending a TRYING message back to the IAD/phone A.
7. After the proxy has validated the request and performed a routing function on the called number, it forwards the INVITE request to IAD/phone B.

8. Once IAD/phone B receives the INVITE, it acknowledges the fact that it received the request by sending a TRYING message back to the proxy server.
9. After IAD/phone B has validated that it can honor the INVITE request, it rings the phone and sends an alerting message to the proxy to indicate this fact. The alerting message can contain the answer SDP. Note that the answer SDP can alternatively be provided in the 200 OK shown in message number 11. Also note if the IAD/phone B could not honor the INVITE request, it would respond here with the most appropriate error response (4xx, 5xx, 6xx).
10. When the proxy server receives the alerting message, it forwards it back to IAD/phone A, which applies a ringback tone to user A, indicating that the far end phone is ringing. The proxy typically has a timer running, and if the user doesn't answer during alerting, it will take some action based on the provisioning information for user B. For example, the call could be forwarded to voicemail as a result of no one answering the call.
11. At some point user B answers the call by going off hook. The IAD/phone B will send a 200 OK response message. Note this is a response message to the original INVITE transaction. The 200 OK response message will contain the answer SDP if the alerting message did not carry this information.
12. When the proxy receives the 200 OK, it forwards it back to IAD/phone A.
13. Now IAD/phone A has the IAD/phone B's answer SDP. It acknowledges the fact that it's ready to send and receive RTP packets by sending an ACK message to the proxy. At this point IAD/phone A considers the call active and cuts through its headset to the RTP stream.
14. The proxy forwards the ACK to IAD/phone B. At this point, the proxy server considers the call active.
15. When the IAD/phone B receives the ACK message, it treats the call as active and cuts through its headset to the RTP stream. Throughout the life of the conversation, both IADs/phones send and receive RTP packets that carry the voice conversation between user A and user B.
16. At some point one of the users decides to end the call by going on hook. In this example user B hangs up, causing a SIP BYE message to be sent to the proxy server. IAD/phone B stops sending and receiving RTP packets, since the call is being released.

17. The proxy server forwards the BYE message to IAD/phone A.
18. When IAD/phone A receives the BYE, it stops sending and receiving RTP packets, since the call is being torn down. It usually applies a tone to the headset, indicating that the far end has terminated the call. The IAD/phone A sends a ACK response message to the BYE. At this point the IAD/phone B considers the call released.
19. The proxy forwards the ACK response to IAD/phone B. At this point the call is completely released.

5.2.6 Session Description Protocol (SDP)

The IETF defined the SDP (Session Description Protocol) as a key signaling protocol component that allows for a generic way to describe multimedia connections over an IP network. SDP provides a mechanism to fully specify the type of multimedia connection to be established. SDP is used consistently across most VoIP protocols, including SIP, H.248, and MGCP. Having a common way to convey multimedia connections allows for easier interworking among network elements in a hybrid VoIP protocol network.

The original SDP RFC 2327 [11] was published by IETF in April 1998. In July 2006, RFC 4566 [12] was published with updates to the SDP specification. SDP defines characteristics of codec configurations for devices that stream IP packets for various types of multimedia applications. Streams can be between two devices or a multicast stream whereby there is one source and multiple receivers. Also supported are conferencing of streams.

Although the *P* in the SDP stands for protocol, it is more of a multimedia description without the message exchange found in most protocols. SDP uses a simple approach to describing connections through a list of name and value pair strings. The names are typically a single character in order to minimize the messaging overhead. A SDP specification is composed of two parts. The first part is the descriptors for the session section, and the second part consists of 0 or more media-level descriptors. Figure 5-9 shows the relationship of the SDP specification and how it relates to a media stream. Stream 1's parameters are fully specified by the session descriptors combined with media 1 descriptors. Stream *N*'s parameters are fully specified by the session descriptors combined with media *N* descriptors.

A *descriptor* provides information about the stream to be used. Multiple streams can also be described by having multiple media-level descriptors. The descriptors in the session's section apply to all streams. Descriptors in the media section apply strictly to the individual stream. If the same descriptor is provided in the media section that exists in the session section, the media section descriptor overrides the session descriptor for that specific media level stream.

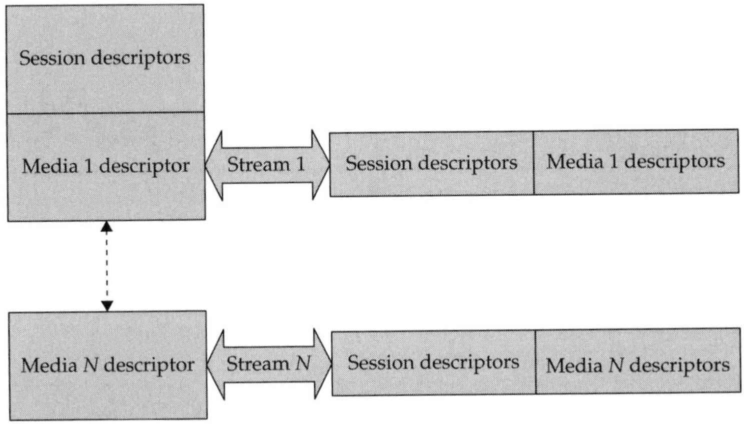

FIGURE 5-9 SDP session media relationship

The following is a list of session-level descriptors with a brief summary of how they are used.

Session Name	Session Value Description
v	Protocol version of the SDP protocol. Currently version 0 is used.
o	This value provides the unique originator application. It consists of a user name, a unique session identifier, and the address of the originating machine.
s	Session name is a text name of the application using the SDP. There should only be one session name.
i	This field provides a text description about the session. It usually defines the purpose of the session.
u	This field is optional and provides a URI link to a web site that can describe more information about the session.
e	This is the contact e-mail address for the coordinator of the session. This field is optional and is usually not populated.
p	This is the session coordinator's phone number. This field is optional and is usually not populated.
c	This value provides the connection information. This is typically an IP address. For each session and media section pair, there needs to be a connection value. The media section connection information overrides the session connection information.
t	Time Descriptor is used to indicate the intended start and stop times of the stream.
r	The repeat descriptor is used in conjunction with the t descriptor for streams that repeat at fixed intervals.

The following is a list of media-level descriptors with a brief description of how they are used.

Media Descriptor Name	Media Descriptor Value Description
m	Each SDP description has one or more media descriptors. Each media descriptor starts with an m line and ends at the next m line or if the end of the SDP is reached. The m descriptor provides the media type being used (audio, video . . .), the protocol being used (RTP, UDP . . .) with a specific format, and the port the packets will be received on.
a	This value provides added attributes about the SDP. This attribute allows implementations to define new attributes as needed. Some examples of common attributes are • **ptime** Defines the packetization time. • **rtpmap** Defines the RTP parameters, including the codec type and clock rate. • **recvonly, sendrecv, sendonly, inactive** These attributes define the send and receiving modes. • **fmtp** This media attribute allows for additional format specifications.
c	This value provides the connection information. This is the same as the c in the session and is optional at media level if a c descriptor exists at the session level. This c value overrides any c value at the session level.
i	This field provides a text description about the media.
b	This optional field provides bandwidth needs for the stream.

5.2.7 SIP Offer/Answer Model

For any multimedia application the parties involved need to agree and coordinate the media streams they send and receive to each other. The SDP RFC, described earlier, provides a mechanism to specify stream characteristics for any multimedia application. However, SDP is a media characterization and doesn't impose any requirements on the exchange of SDP or the details on how pieces of equipment converge to a final agreement of the streams they will use. To address these issues, the IETF defined RFC 3264 [13], which is known as the offer/answer model.

The offer/answer model provides guidelines on how to establish and modify streams among various applications. The model defines an application-level exchange of media information specified via SDP that can be used across VoIP signaling protocols such as SIP, MGCP, and H.248. The offer/answer model is intended to be protocol independent so that it can be applied to any VoIP protocol.

The originator of a multimedia session sends out an initial offer toward the destination. In SIP this would be the INVITE message (message 5) shown in Figure 5-8. This initial offer will contain the description of each proposed media stream with options for all codecs supported. For each stream there is a dedicated m line listing the different media formats supported. If RTP is being used, an "a=rtpmap" mapping attribute must be present indicating the supported codecs. Either unicast or multicast streams can be specified by each SDP "m=" line. What follows is an example of using multiple streams by an application that requires both audio and video. Included in the initial offer is the mode to start with, which could be receive only, send only, or both send and receive. The following is an example of an initial offer:

```
v=0
o=femtocellphoneA 14360 15217 IN IP4 10.10.180.111
s=Offer Example
t=0 0
c=IN IP4 10.1.2.3
m=audio 12345 RTP/AVP 0 8 18
a=rtpmap:0 PCMU/8000
a=rtpmap:8 PCMA/8000
a=rtpmap:18 G729/8000
m=video 67890 RTP/AVP 11 14
a=rtpmap:11 H261/90000
a=rtpmap:14 H263/90000
```

This example shows the offer proposing an audio media stream with three choices of audio codecs, given in descending order of preference as PCMU, PCMA, and G729. These three choices are designated by the 12345 audio line listing the corresponding rtpmap line identifiers (0 8 18). The second media stream proposes a video stream with two choices of video codecs given in descending order of preference as H.261 and H.263. Similarly the two choices are designated by the 67890 video line listing the corresponding rtpmap line identifiers (11 14).

Once the destination subscriber's equipment receives the offer, it will choose the codec to use for both the audio and the video streams. The following is an example answer SDP:

```
v=0
o=femtocellphoneB 1432 12336 IN IP4 10.160.10.111
s=Answer Example
t=0 0
c=IN IP4 10.2.3.4
m=audio 54321 RTP/AVP 0
```

```
a=rtpmap:0 PCMU/8000
a=sendrecv
m=video 67890 RTP/AVP 14
a=rtpmap:14 H263/90000
a=sendrecv
```

This answer shows that femtocellphoneB chooses the first audio codec that uses PCMU and sets its mode to send and receive. Also chosen is the second proposed video codec that uses H.263 and sets its mode to send and receive. With having both the audio and video as send and receive, the application here would be for a two-way video phone call.

If a device involved in a multimedia session decides to remove one of its streams, it can do so by sending an offer similar to the original offer except that the stream being removed has its port set to 0. Also the stream being subtracted only needs to list its m line. The following shows an offer sent from femtocellphoneB dropping the video stream, thereby reducing the video call to an audio-only call:

```
v=0
o=femtocellphoneB 1432 12336 IN IP4 10.160.10.111
s=Answer Example
t=0 0
c=IN IP4 10.2.3.4
m=audio 54321 RTP/AVP 0
a=rtpmap:09876 PCMU/8000
a=sendrecv
m=video 0 RTP/AVP 14
```

If a device involved in a multimedia session decides to modify a stream, it can do so by simply sending an offer similar to the last exchanged offer except with the proposed changes to the stream being modified. If it is desired to add a stream, then an offer should be sent with the media proposal being added to the end of the existing SDP. As a result, the peer will reply with a corresponding answer SDP. The following shows an offer sent from femtocellphoneB with the video stream being proposed to be added back in again, using a H263 video codec this time:

```
v=0
o=femtocellphoneB 1432 12336 IN IP4 10.160.10.111
s=Answer Example
t=0 0
c=IN IP4 10.2.3.4
m=audio 54321 RTP/AVP 0
a=rtpmap:0 PCMU/8000
a=sendrecv
m=video 51372 RTP/AVP 99
a=rtpmap:99 h263-1998/90
a=sendrecv
```

All of the offer/answer SDP exchanges are carried within the body of a VoIP signaling message. As mentioned earlier, the initial offer would be sent in an outgoing INVITE. Offers can be answered in one of many end-to-end SIP messages, which include 18X and 200 OK. If an offer is being rejected, a SIP rejection message can be sent such as 4XX. Modifications, including addition and deletion of streams, for an active multimedia session can be achieved by sending a RE-INVITE with the modified offer SDP for all active calls. These RE-INVITE offers are answered with 200 OK, since the call is already active.

References

[1] January 2003, IETF RFC 3435 "Media Gateway Control Protocol (MGCP) Version 1.0"; F. Andreasen, B. Foster.

[2] October 1999, IETF RFC 2705 "Media Gateway Control Protocol (MGCP) Version 1.0"; M. Arango, A. Dugan, I. Elliott, C. Huitema, S. Pickett.

[3] June 2002, IETF RFC 3261, "SIP: Session Initiation Protocol"; J. Rosenberg, H. Schulzrinne, G. Camarillo, A. Johnston, J. Peterson, R. Sparks, M. Handley, E. Schooler.

[4] March 1999, IETF 2543,"SIP: Session Initiation Protocol"; H. Schulzrinne, M. Handley, J. Rosenberg, E. Schooler.

[5] June 2002, IETF 3262, "Reliability of Provisional Responses in the Session Initiation Protocol (SIP)"; J. Rosenberg, H. Schulzrinne.

[6] June 2002, IETF 3265, "Session Initiation Protocol (SIP)-Specific Event Notification"; A.B. Roach.

[7] October 2004, IETF RFC 3903, "Session Initiation Protocol (SIP) Extension for Event State Publication"; A. Niemi.

[8] October 2000, IETF RFC 2976, "The SIP INFO Method "; S. Donovan.

[9] April 2003, IETF RFC 3515, "The Session Initiation Protocol (SIP) Refer Method"; R. Sparks.

[10] September 2002, IETF RFC 3311, "The Session Initiation Protocol (SIP) UPDATE Method"; J. Rosenberg.

[11] April 1998, IETF RFC 2327, "SDP: Session Description Protocol"; M. Handley, V. Jacobson.

[12] July 2006, IETF RFC 4566, "SDP: Session Description Protocol"; M. Handley, V. Jacobson, C. Perkins.

[13] June 2002, IETF RFC 3264, "An Offer/Answer Model with the Session Description Protocol (SDP)"; J. Rosenberg, H. Schulzrinne.

CHAPTER 6

Media Protocols over IP

IP networks were originally designed to transport data exchanged between computers. A typical application is transferring files from one computer to another. IP networks use packet switching technology that is very efficient at transferring bursty traffic at a high throughput rate. However, the initial IP networks experienced significantly high end-to-end delays, which were acceptable for non-real-time data applications but not suitable for real-time applications such as voice and video transmission. As IP networks evolved and their performance improved, the end-to-end delay decreased dramatically, allowing the quality of service needed by real-time applications. In order to transfer a real-time signal, such as analog voice, over a digital network, an electronic device known as a codec is required. This chapter explores the details of various types of codecs and the protocols used to transport real-time traffic over an IP network.

6.1 Voice Codecs

A voice codec converts a person's speech into a digital encoding that is transported over a digital network. On the far end of the digital network, a voice codec is used to convert the digital voice encoding back to an analog signal that is played out as voice using a speaker, ear piece, or handset.

Figure 6-1 shows a block diagram of the basic components of a voice codec. On the right side of the diagram are the analog voice interfaces, which include a handset, a microphone in, and a speaker/handset out. The hybrid splits the two-wire interface into a four-wire interface.

The user speaks into the microphone or the headset, where it is transformed into an analog signal. The microphone in and the transmit signal of the handset are summed together and fed into the analog-to-digital converter (ADC). The output of the ADC is fed into

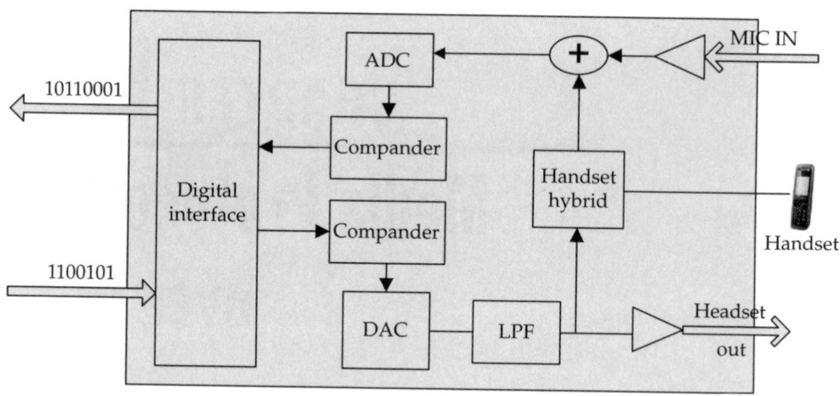

FIGURE 6-1 Voice codec block diagram

a digital compander to increase the sender's dynamic range. A description of voice dynamic range is provided in the G.711 section [1]. The compander mode is typically programmable, such as either A-Law or µ-Law. The companded transmit signal is fed into the digital interface logic, where the digitized voice is transmitted. The received digital signal is fed to the receiver's compander, where it is uncompressed and fed into the digital-to-analog converter (DAC). The analog output of the DAC is fed into a low-pass filter (LPF). The LPF is used to remove any high-frequency signals that have occurred as a result of quantization. The output of the low-pass filter is fed into the handset hybrid and the headset output, where the far-end voice can be heard.

6.1.1 PCM (Pulse Code Modulation)

Voice codecs were first used in telephone networks to convert analog voice conversations to a digitally encoded signal. Conventional TDM voice networks will digitize an analog voice signal and place it into a digital time slot channel. Once in a time slot, the voice conversation can be switched and routed to a final destination where a codec will convert the time slot back into an analog signal. TDM-based telephone networks use an encoding scheme called pulse code modulation (PCM). A PCM codec samples an analog signal 8000 times a second and performs an A/D (analog-to-digital) conversion. Public telephone networks have been designed to have a 4 KHz (kilohertz) bandwidth, which provides reasonable voice quality. The 8000 PCM samples per second value is derived from the Nyquest rate, where the sampling should be done at a level at least twice the bandwidth being digitized. For each analog sample, an eight-bit number reflecting its amplitude is assigned, which is referred to as its quantization value. Figure 6-2 provides an illustration of a five-bit PCM quantization of a sine wave. A few sample digital values shown in Figure 6-2 are

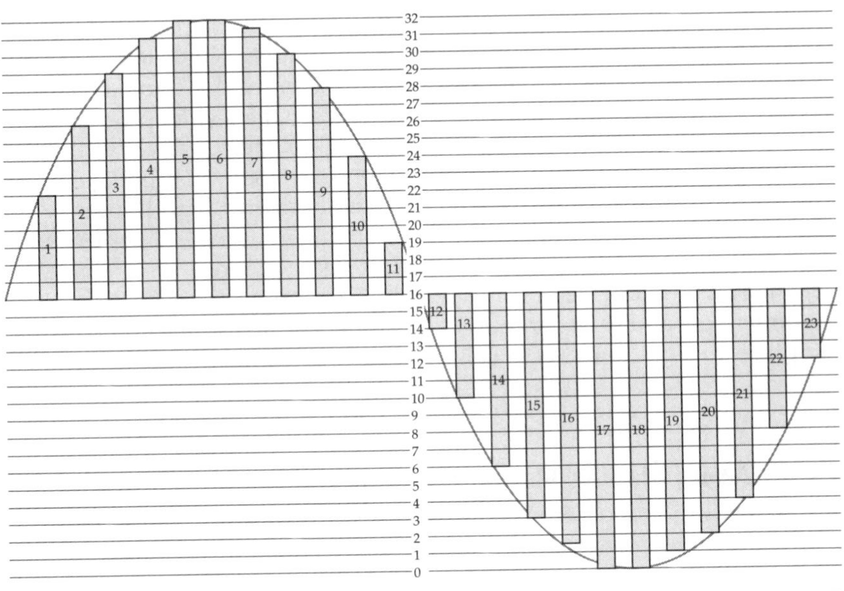

FIGURE 6-2 5 BIT PCM signal quantization

sample 1 -> 22, sample 6 -> 31, sample 12 -> 14, sample 18 -> 0, and sample 23 -> 12.

On the encoder or modulation side an ADC is used to digitize the analog signal. On the far end the digitized signal is converted back into a replication of the original voice signal using the reverse process of modulation, which is called *demodulation.* Each eight-bit digital sample received is converted back into an analog voltage value reflecting the digital value received. This analog value is held until the next eight-bit sample is received, when the voltage is updated to reflect the new value. On the decoder or demodulation side an ADC is used to convert the digital PCM back to an analog signal.

6.1.2 G.711 A-Law and µ-Law Companding

Human hearing can distinguish sound that varies in volume from a tiny whisper to a loud musical band. This range of sound detection is referred to as the sound dynamic range. Similarly, human vision also has a wide dynamic range for light intensity. When digitizing a signal, very small changes in a low-volume signal tend to be lost as the digital value representation approaches a single-bit change. For example, an eight-bit PCM signal has a resolution of 1/256, since eight bits can represent 256 different values. The error in digitizing a signal is called the *quantization error.* To compensate for quantization errors, a process called *companding* is performed. Companding increases the amplitude of low-volume sounds in what is called expansion. Conversely, companding decreases the amplitude of high-volume sounds in what is called compression. Companding is performed on an analog signal just prior to the analog-to-digital conversion. Then the reverse process is performed on the far end just after the digital-to-analog conversion. Two different companding algorithms widely used today in telephone networks are known as A-Law and μ-Law. The μ-Law algorithm is used in North America and Japan. A-Law is used in Europe and other parts of the world. A-Law is very similar to μ-Law; however, it was designed for easier processing by computers directly.

Both μ-law and A-law encode 14-bit and 13-bit signed linear PCM samples into 8-bit chunks after companding is performed. Since PCM uses a sample rate of 8000 samples per second, it results in a 64 Kbps data stream.

6.1.3 AMR (Adaptive Multirate Compression)

Adaptive multirate compression (AMR), as the name implies, is a codec that not only compresses the bandwidth required but also can adapt to one of multiple rates. AMR was designed for wireless networks, where bandwidth is a premium and the quality of the connection changes during the life of the call. 3GPP incorporated AMR into GSM and UMTS specifications based on its benefits. It is a widely used codec, since it works extremely well with wireless technologies. AMR is required to be supported by IMS terminals based on the 3GPP standard. AMR uses 20 millisecond frames and varies the number of bits per frame based on the mode. Table 6-1 shows different modes, their bit rates, and the number of bits used per frame.

When the signal-to-noise (S/N) ratio is high, the mode is set to a higher rate such as AMR 12.2. As the S/N ratio decreases for various reasons, including radio interference, AMR adjusts its rate down to a mode closer to AMR 4.75. AMR has the ability to change rates from one millisecond to the next, although shifting rates too quickly may not be supported by some networks.

AMR 12.2 is also known as ETSI GSM Enhanced Full Rate (GSM-EFR), which was used to significantly improve the voice quality for GSM Full Rate. The GSM-EFR also supports both A-Law and μ-Law as well as conversions between the two. AMR 7.4 is compatible with TDMA Enhanced Full Rate used by TDMA networks to improve voice quality. AMR 6.7 is used in Japan and is compatible with 2G Personal Digital Cellular Enhanced Full Rate (PD-EFR).

6.1.4 AMR-WB (Adaptive Multirate Compression Wide Band)

AMR-WB (Adaptive Multirate Compression Wide Band) samples voice at 16,000 times per second, which is twice the rate of conventional TDM voice networks. Doubling the number of samples dramatically improves the voice quality. The voice bandwidth range is

Mode	Bits per Frame	Full Rate/ Half Rate	Bit Rate (Kbps)	Compatible Standards
AMR 12.2	244	Full Rate	12.2 Kbps	ETSI GSM Enhanced full Rate
AMR 10.2	204	Full Rate	10.2 Kbps	
AMR 7.95	159	Full Rate/ Half Rate	7.95 Kbps	
AMR 7.4	148	Full Rate/ Half Rate	7.4 Kbps	TDMA-EFR
AMR 6.7	134	Full Rate/ Half Rate	6.7 Kbps	PDC-EFR
AMR 5.9	118	Full Rate/ Half Rate	5.9 Kbps	
AMR_5.15	103	Full Rate/ Half Rate	5.15 Kbps	
AMR_4.75	95	Full Rate/ Half Rate	4.75 Kbps	

TABLE 6-1 AMR Modes

increased to 50–7000 Hz compared to 300–3400 Hz using PCM encoding. This bandwidth range increase provides AMR-WB with the ability to transport stereo-quality sound. AMR-WB is also capable of adjusting its rate based on the quality of the connection similar to AMR using nine different bit rates. The lower rates used by AMR-WB are 6.6 Kbps, 8.85 Kbps, and 12.65 Kbps. These lower rates are only used if the radio connection is poor. The higher rates provide superior sound qualities, which are 14.25 Kbps, 15.85 Kbps, 18.25 Kbps, 19.85 Kbps, 23.05 Kbps, and 23.85 Kbps. AMR-WB is specified in 3GPP IMS for wideband services.

6.1.5 G.718

G.718 is a codec defined by ITU and can be adjusted to one of five different rates of 8, 12, 16, 24, and 32 Kbps per second [2]. At low data rates, the codec provides decent-quality audio with a bandwidth of 250 Hz to 3.5 KHz. At the higher rates, the codec provides high-quality voice with a bandwidth of 50 Hz to 7 KHz. G.718 is very resistant to frame errors as a result of transmission errors, providing minimal human detection of voice quality degradation. This robustness allows G.718 to be used in many different applications, including IP access for fixed, wireless, and mobile networks. The codec advances used in G.718 provide the quality of G.722 at a much lower rate of 12.65 Kbps [3]. It also provides decent-quality speech equivalent to G.729 at 11.8 Kbps [4]. The G.718 embedded frame structure allows network elements the ability to downgrade the quality and reduce the data rate if needed. The embedded frame structure is also scalable for future higher rates, allowing stereo quality bandwidth of 50 Hz to 14 KHz. G.718 uses discontinuous transmission (DTX) techniques to optimize the efficiency of the communications channel. DTX is an algorithm that only transmits when the end user is speaking. Statistically during a phone conversation, someone is speaking only half the time, and resources can be conserved, especially power consumption, if the transmitter is turned on only when speech is present. Along with DTX, G.718 uses comfort noise tone generation (CNG) on the receiver end, providing a very low-volume background noise when the transmitter is not sending anything as a result of no conversation. This background or comfort noise is used to prevent callers from observing complete silence from the effects of the DTX. Without the CNG, callers would get the sensation that their connection had been lost, due to the complete silence.

6.1.6 G.721

G.721 is a codec specified by ITU that uses adaptive differential pulse-code modulation (ADPCM) technology at speeds of 16, 32, 48, 56, and 64 Kbps [5]. ADPCM adjusts the amount of quantization performed. For example at 64 Kbps, ADPCM uses the standard PCM eight bits. The most popular ADPCM rate is 32 Kbps, which is achieved by using one less bit of encoding (seven bits) to realize half the bandwidth of PCM. As the rate decreases and fewer bits are used for encoding, the voice quality diminishes.

6.1.7 G.722

G.722 is a codec specified by ITU for use at lower rates. It uses adaptive multirate wideband (AMR-WB), where it adjusts its rate to suit the quality of the connection. The rates supported are 48, 56, and 64 Kbps. G.722 samples voice 16,000 times per second, which is twice the rate of conventional TDM voice networks. Doubling the number of samples dramatically improves the voice quality. At higher rates, G.722.2 uses AMR-WB for superior sound quality.

6.2 Video Codecs

A video codec converts a video signal into a digital encoding that can be transported over a digital network or stored to a device like a disk drive or flash memory. On the far end of the digital network, the video codec converts the digital encoding signal back to a video signal that can displayed on a screen, a computer monitor, or even a mobile phone display. A video signal is encoded by taking repetitive snapshots as still images and encoding each snapshot. Shortening the interval between snapshots results in a higher-quality video signal replication. The encoded video signal can be played back by decoding each snapshot repetitively at the same interval the video source was encoded with. An image is constructed on a display using a collection of what are commonly known as pixels. A *pixel* is the smallest component that makes up a digital display. Figure 6-3 shows a sample diagram of a digital still image using 20 × 15 resolution of pixels. There are 15 rows and 20 columns that make up this pixel matrix.

When encoding, the still image information about each pixel is captured, including its combination of colors (red, green, blue) and light intensity. The number of colors each pixel can represent is referred to as its *color depth.* The greater the color depth, the more bits are needed per pixel. For example, 1 bit would be 2 colors (black or white), whereas 8 bits could represent 256 different colors. Computers and TV monitors are rated by their resolution or number of pixels supported. The larger the number of pixels a display uses, the better the display quality. Table 6-2 shows the various computer video standards and their related pixel resolution attributes.

An SVGA screen using a 16-bit color depth would require (16 × 800 × 600) = 76.8 megabits for one image frame. A standard rate of 30 frames per second would require 76.8 megabits × 30 = 2304 megabits per second of bandwidth to transport this video signal. This is a huge amount of bandwidth needed to transport an average-quality computer's video.

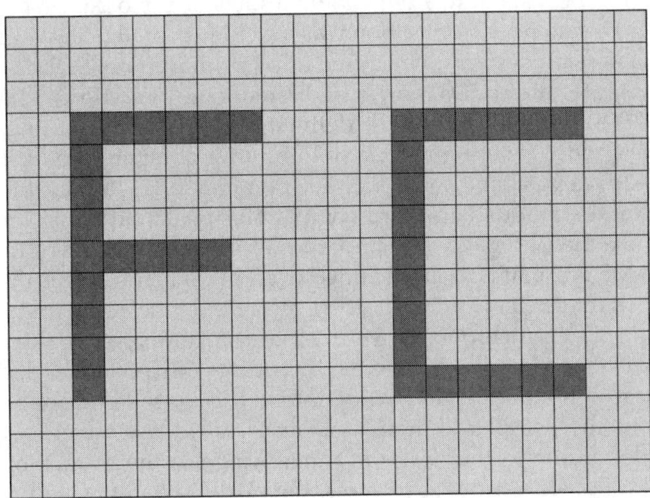

FIGURE 6-3 A 20 × 15 pixel display

Computer Video Standard	Width × Height	Resolution in Megapixels
CGA	320 × 200	0.064
EGA	640 × 350	0.224
VGA	640 × 480	0.3
SVGA	800 × 600	0.5
XGA	1024 × 768	0.8
SXGA	1280 × 1024	1.3
UXGA	1600 × 1200	1.9
WUXGA	1900 × 1200	2.3

Table 6-2 Standard Computer Video Resolutions

Transporting large data streams like this is not cost effective, and as a result most if not all video codecs today use compression technology to reduce the overall bandwidth required. Digital video codecs are used in many applications, including mobile phones, DVD players, computers, and broadband TV set top boxes. For Internet applications, various types of codecs are used, which many times requires the user to download a different codec player to view videos.

6.2.1 MPEG

The Motion Picture Experts Group (MPEG) working group defined a set of standards for video and audio formats to be used for video conferencing and media storage. The first standard they defined was MPEG-1 to be used at relatively low bit rates of 1.5 to 2 megabits per second. The intended application was for video conferencing over T1/E1 transmission systems, which included primary rate ISDN. At this low data rate, audio CDs can be used for storage of a MPEG-1 file.

MPEG-2 was subsequently defined for applications using full-broadcast video, at a higher bandwidth range compared to MPEG-1. MPEG-2 was designed to operate in the range of 3–15 megabits per second and supports backward compatibility with MPEG-1. Today's DVDs use the MPEG-2 video and audio format. For audio, the well-known MP3 format is used. Included in all MPEG encoding is the system clock reference, which is used to reconstruct the audio and video to be properly synchronized. MPEG-2 uses a compression algorithm that reduces the bandwidth needed by a factor of 15–30 times and still provides high-quality standard-definition video. Video compression partitions the image into groups of pixels called macroblocks. From one video frame to the next, only the change in the macroblocks is encoded. For slow-moving pictures, the picture quality is very good. With sports and quick-moving scenes, picture artifacts can be seen.

As MPEG-2 matured, it was enhanced to support high rates, allowing for High Definition Television (HDTV) support. H.262 is a codec defined by ITU that essentially follows the MPEG-2 standard.

MPEG-4 has been built on many of the advantages of MPEG-1 and MPEG-2 while extending their capabilities to allow for more multimedia applications. MPEG-4 defines software and hardware objects that allow advanced applications to be developed for orchestrating video, audio, and graphic animations. A large and full-featured set of tools for encoding is available with so many choices that subsets of features are defined called *profiles*. MPEG-4's capabilities are still evolving as a result of the many capabilities it supports requiring vendor interoperability.

6.2.2 H.261, H.262, H.263

H.261 was originally designed to be used over circuit-switched channels of an ISDN line at multiples of 56 or 64 Kbps [6]. The video bit rate range is from 40 Kbps to 2 Mbps and is used primarily in older videoconferencing and video phone products. Although H.261 is not widely used today, many codecs that followed leveraged off of its encoding techniques. Macroblocks of 16 × 16 pixels are defined in H.261 as part of its compression algorithm. Only changes to a macroblock need to be encoded from one image frame to the next, thereby significantly reducing the amount of information needed to be encoded. H.120 was the predecessor to H.261, which offered significant improvements [7].

ITU's Video Coding Experts Group (VCEG) and the ISO MPEG jointly developed the H.262 standard [8]. Essentially, H.262 and MPEG-2 are identical specifications, apart from the labeling with the standards body numbering.

H.263 is a codec initially defined by ITU for purposes of low-bit-rate video conferencing [9]. It was also defined by ITU VCEG (Video Coding Expert Group) in the 1995 and 1996 time frame. Original users of H.263 were on circuit-switched networks for video conferencing. As time progressed and Internet access grew, H.263 became widely used on IP videoconferencing using RTP. Several popular free video sharing web sites have delivered video streams using H.263. The 3GPP IMS specification references H.263 as a required codec to be supported for its Multimedia Messaging Service (MMS).

H.264, like other codecs, was jointly defined by VCEG and MPEG [10]. Again as a result, H.264 and MPEG-4 are technically equivalent standards. The original goal of H.264 was to provide a decent-quality video codec with significantly lower bit rates in comparison to other existing codecs. H.264 was also designed to scale up to higher bit rates, offering high resolution video when broadband IP access is available. In order to scale to different rates, multiple types of codec encodings are used. The standard defines profiles for each type of encoding supported, allowing implementations to choose from all or a subset to support.

As with the other H-series codecs, H.264 capabilities evolved over time. Extensions were added to support higher-resolution video by using a technology called Fidelity Range Extensions (FRext), increasing the sample bit depth precision. Annex G of the standard defines a scalable video coding that allows for video sources to take advantage of bit rate peeling. *Bit rate peeling* is a mechanism where a video is encoded and stored at a high-quality bit rate and then can be streamed to different destinations at different rates, depending on the network being used. This is extremely useful in wireless and IP networks, where the end-to-end network bit rate can vary dramatically, without requiring the video source to store multiple video encodings. As a network rate is determined, the source of the video stream will adjust down the available bit rate while providing the best possible video quality.

6.2.3 Video Codec Quality

There are several codec encoding parameters that have trade-offs between video picture quality robustness and available data rate. When a codec uses pixel groups or blocks like in MPEG 2, the larger the block size, the lower the data rate that is needed. When a block error occurs, however, it takes longer for the video to recover, making the block error more noticeable to the end user. Another set of parameters used by codecs are the quantization coefficients. The larger or more coarse the coefficients chosen, the greater the impact on the picture distortion due to higher quantization error. The frame rate parameter also significantly affects the dynamic behavior of the codec. As seen earlier, a typical frame rate ranges from 30 to 60 frames per second. Lowering the frame rate will reduce the data rate needed; however, it will degrade the picture quality, since the video image is being refreshed less often, making movement in the video much more distorted.

6.3 Text Encoding

Today sending text messages with mobile devices is used as much if not more than voice phone calls. Text applications such as text messaging, news alerts, and financial information use one of several text technologies. In general, texting uses dramatically less network resources compared to real-time applications like voice and video, primarily because the data rate needed is extremely reduced for texting. Text applications do not require a codec, since information is captured directly to a digital format, so no analog-to-digital conversion is needed. To send one character in text without compression takes only seven bits using ASCII (American Standard Code for Information Interchange).

6.3.1 Short Message Service

The most widely used text or instant messaging technology is the Short Message Service (SMS) protocol. This protocol was originally designed for GSM wireless devices and has been adopted by CDMA, satellite, and fixed land networks. SMS allows wireless devices to send text messages to other wireless devices or software applications such as an Instant Messaging Web Service. SMS is supported by legacy telephone SS7 networks using the Mobile Application Part (MAP). The CAMEL (Customized Applications for Mobile Networks Enhanced Logic) protocol has been extended to support SMS allowing IN (Intelligent Network) text messaging applications to be offered. Today text messaging services generate billions of dollars of revenue, with SMS being the most widely used. Other protocols have also been developed for the purposes of texting. Many proprietary protocols have been developed, especially using IP. Standard SMTP (Simple Mail Transfer Protocol), which is used for e-mail, has also been used to offer text messaging.

6.3.2 Instant Messaging

Instant messaging (IM), just like mobile phone texting, is an extremely popular application used on the Internet today. Instant messaging allows two or more users to exchange content, typically text, in pseudo-real time. IM is not quite as real time, unlike ToIP, which transmits characters as soon as they are typed. With IM, a message will only be delivered to a recipient after the sender explicitly presses Send or the Enter button. Besides text, other IM contents can be a picture, a sound or video clip, or even an attached file.

RFC 2778 [11] defines the presence and instant messaging model. The presence capability allows users to know the status of users—whether they are online and ready to accept an instant message. Many services will have a friend list where permission needs to be granted for a user's presence to be made available to another particular user. There are many vendor implementations of instant messaging. RFC 3860 [12], called Common Profile for Instant Messaging (CPIM), was written to allow vendors to capture the characteristics of their IM implementation, which assists in the development of IM gateways, thereby promoting IM service interoperability. CPIM specifies having instant messages follow a standard content format known as Multipurpose Internet Mail Extensions (MIME). RFC 3862 [13] specifies how MIME is used to be compliant with CPIM, including defining the from header, the to header, a cc header, and date and time headers.

Similar to SIP, a URI is used to address an IM mailbox, and its complete syntax is specified in RFC 2368 [14]. An example of an IM URI would be im:mike@femtocell.com. There exist two modes of instant messaging, called pager-mode and session-based instant messaging. The pager-mode instant messaging sends a message independent of any other instant message without requiring any preestablished session. The pager-mode name reflects the fact that pagers also send a single message without any relationship to other pager messages. SIP defined a command specifically for IM called MESSAGE. A SIP user agent client (UAC) sends an IM message by encapsulating the text message into a SIP message command. When a UAC receives a 200 OK response to the message, it considers it delivered. Although the message was delivered, that doesn't necessarily mean it has been read. Some services will store the message for retrieval at a later time when the recipient is logged in. Figure 6-4 from RFC 3428 [15] shows a pager-mode instant message exchange using the new SIP MESSAGE command.

6.3.3 MSRP

The Message Session Relay Protocol (MSRP), as the name implies, is a session-based transport protocol to exchange instant messages. An SDP (Session Description Protocol) offer/answer is exchanged to establish a MSRP session in order to correlate all instant messages between two users. Using SDP offer/answer enables the advantage of including additional media with the MSRP dialog such as voice, video, or an image. In comparison to the pager-mode IM, a session-based IM would be useful if a significant exchange of IM messages is expected between two or more IM participants. One of the most commonly used protocols to exchange offer/answer SDP is SIP. RFC 4975 [16] defines the SIP-based MSRP protocol. An MSRP session is established essentially the same way a SIP voice or video session is established. A SIP client initiates an MSRP session by sending a SIP INVITE with an MSRP offer SDP. Figure 6-5 is an example INVITE with a 200 OK response, similar to an example provided by RFC 4975. The INVITE shows an SDP offer using MSRP over TCP. MSRP is required to use a transport protocol that provides flow control such as TCP. UDP would not be used, since it is best effort only and doesn't provide for flow control, which would cause network congestion when excessive IM traffic occurs. Included in the INVITE SDP is the source IP address and the TCP port to be used to send MSRP messages from Joe's phone.

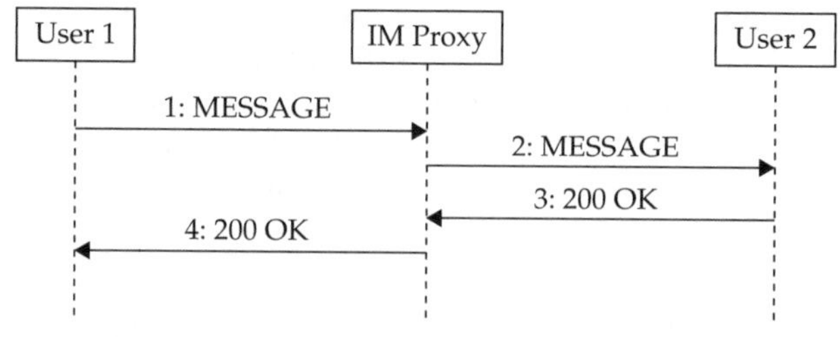

Figure 6-4 SIP IM pager mode

```
INVITE sip:mike@femtocell.phone.com SIP/2.0
         To: <sip: mike@femtocell.phone.com >
         From: <sip:joe@work.phone.com>;tag=541
         Call-ID: 5761zx94KY
         Content-Type: application/sdp
         c=IN IP4 work.phone.com
         m=message 7699 TCP/MSRP *
         a=accept-types:text/plain
         a=path:msrp://work.phone.com:7699/bjfX9vyadqs;tcp
SIP/2.0 200 OK
To: <sip: mike@femtocell.phone.com >;tag=047mr
         From: <sip:joe@work.phone.com>;tag=541
         Call-ID: 5761zx94KY
         Content-Type: application/sdp
         c=IN IP4 femtocell.phone.com
         m=message 12763 TCP/MSRP *
         a=accept-types:text/plain
         a=path:msrp://femtoell.phone.com:24985/cmpd44t6z90x5a;tcp
```

FIGURE 6-5 MSRP INVITE exchange

The 200 OK response includes the destination's IP address and TCP port to be used to send MSRP messages from Mike's phone.

Once the MSRP SIP session is established, MSRP SEND messages are used to exchange messages between users. The content of the MSRP SEND message is the text message entered by the user, usually encoded using MIME [17]. The MSRP SEND message receipt is acknowledged with a 200 OK message. Figure 6-6 shows an MSRP session establishment between phones of two users, Joe and Mike, with a SIP B2BUA (Back to Back User Agent). Messages 1–8 involve the SIP session establishment using an INVITE including an MSRP SDP offer/answer exchange. All session establishment messages are orchestrated through the B2BUA. As part of the session establishment, a MSRP socket port, most likely a TCP port, is opened. Once the session establishment has completed, either user can use an MSRP SEND message to exchange an IM. Each MSRP message is sent directly between the end users, Mike and Joe, and is individually acknowledged with a 200 OK.

6.3.4 Text over IP

Text over IP (ToIP) offers real-time texting between two end users. It utilizes the Real Time Protocol (RTP) to transport text in an IP packet. Two users connected in a ToIP text session can expect real-time text behavior whereby as soon as a character is typed, it will appear on the peer's user display. One major benefit of ToIP using RTP is that it allows its stream timing to be synchronized to other streams that use RTP, such as voice and video. Synchronizing ToIP with other streams can be very useful in screen-sharing applications. RFC 4103 RTP for

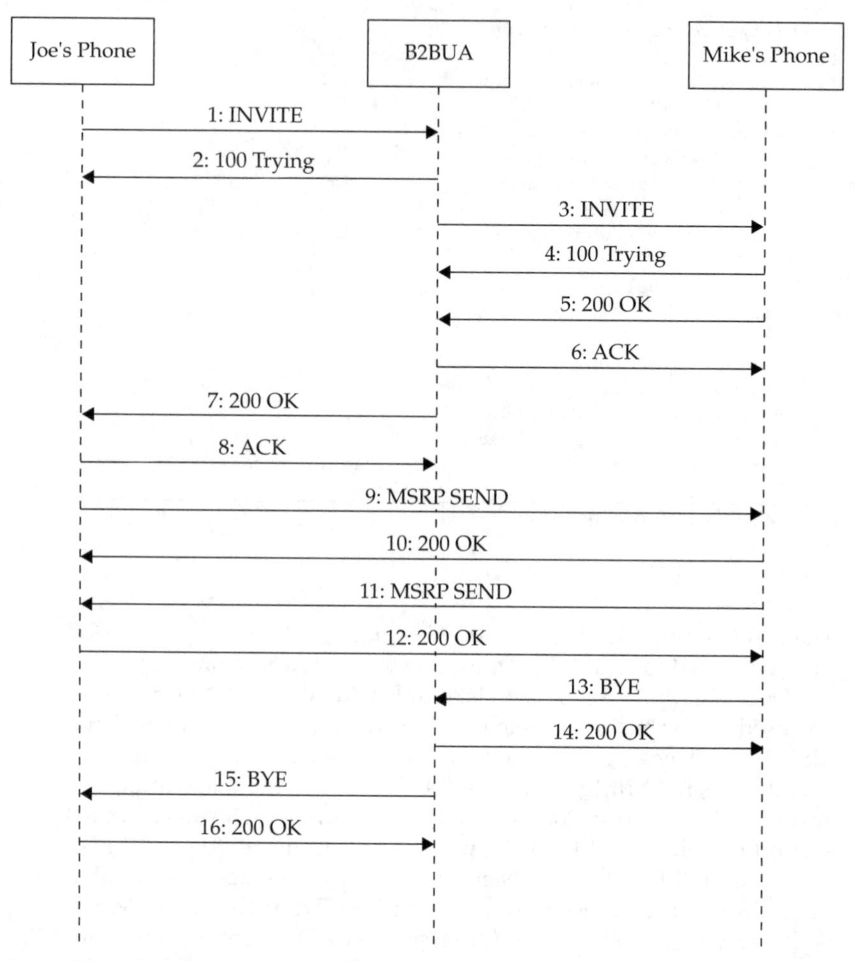

FIGURE 6-6 MSRP session establishment

Text Conversion (RTPTC) [18] defines the RTP formatting for Text over IP conforming to ITU-T Recommendation T.140 [19]. RTPTC uses sequence numbers to allow for proper order of characters typed. In general one character is sent at a time, with an exception of cutting and pasting phrases from the user interface. Also offered for high reliability is a redundancy mechanism whereby duplicate packets are sent to ensure a higher probability of a character being received by the targeted recipient. Figure 6-7 shows the protocol stack for ToIP. The stack consists of the T.140 Real Time Text Presentation Protocol at

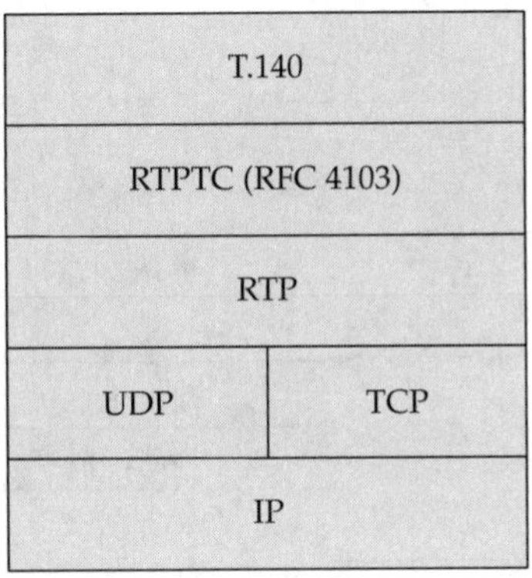

Figure 6-7 ToIP protocol stack

the highest layer, and then RFC 4103 is used to format the text into RTP frames. RTP can be configured to operate over UDP or TCP. Of course, all these protocols are encapsulated into IP packets.

In order to establish the ToIP session again, a SIP SDP offer/answer is exchanged using an INVITE. For comparison purposes, Figure 6-8 shows the same INVITE exchanged as in the MSRP session establishment, with the exception that the SDP offer here is for a ToIP connection. The offer in the INVITE is specifying using port 12000, with T.140 RTP encoding using a 1000 Hz clock frequency.

```
INVITE sip:mike@femtocell.phone.com SIP/2.0
          To: <sip: mike@femtocell.phone.com >
          From: <sip:joe@work.phone.com>;tag=541
          Call-ID: 5761zx94KY
          Content-Type: application/sdp
          c=IN IP4 work.phone.com
          m=text 12000 RTP/AVP 100
            a=rtpmap:100 t140/1000
SIP/2.0 200 OK
To: <sip: mike@femtocell.phone.com >;tag=047mr
          From: <sip:joe@work.phone.com>;tag=541
          Call-ID: 5761zx94KY
          Content-Type: application/sdp
          c=IN IP4 femtocell.phone.com
          m=text 13000 RTP/AVP 101
            a=rtpmap:101 t140/1000
```

Figure 6-8 ToIP INVITE exchange

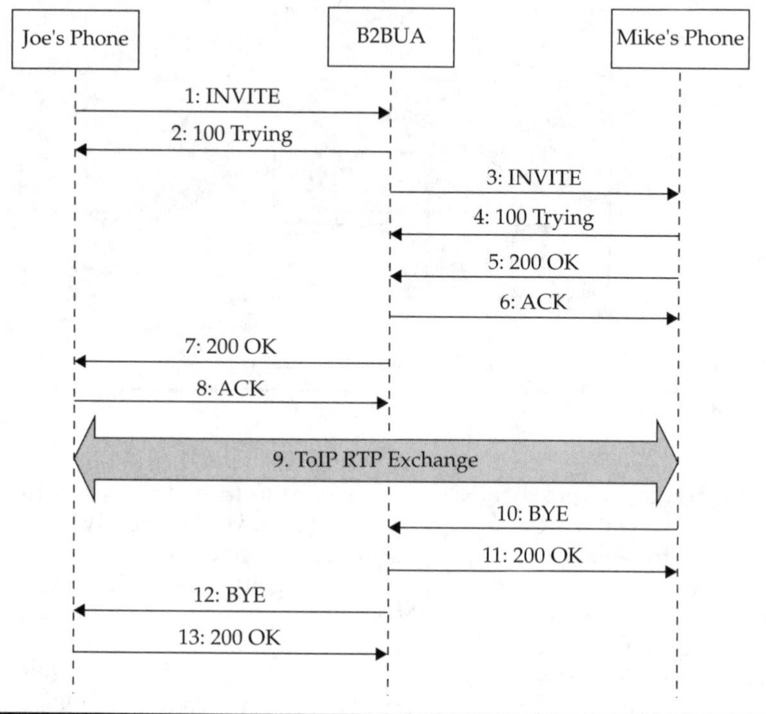

FIGURE 6-9 ToIP session establishment

The session establishment is also similar to that in MSRP. Figure 6-9 shows the ToIP session establishment message exchange. The significant difference here is that once the session has been established, the end-user text is sent over a ToIP RTP connection as opposed to a SIP MSRP SEND message.

6.4 Media Transport Protocols

In order to carry real time media traffic on an IP network a media transport protocol is required. The following sections describe a few of the most commonly used IP media transport protocols.

6.4.1 RTP

The Real-Time Transport Protocol (RTP) was originally defined by RFC 1889 [20] and subsequently updated by RFC 3550 [21]. RTP is most widely used in VoIP networks to transport real-time media such as voice and video. RTP streams are typically established using a VoIP signaling protocol such as MGCP, H.248, H.323, or SIP. The base RTP protocol addresses the common aspects of real-time transmission over an IP network. For each specific codec a profile exists that defines the encoding to be inserted into RTP packets.

Figure 6-10 shows the RTP header used by all RTP packets. The version field (bits 1 and 2) defines the RTP version, which is currently 2. The PT field (bits 9–15) defines the payload type, which corresponds to the type of codec being used. At times the PT value can be changed during the life of a RTP connection for purposes of increasing or decreasing bandwidth based on the quality of the IP connectivity between the two end codecs. The sequence number is a number that increases by 1 for each RTP packet sent. This affords the receiver the ability to know how many packets have been lost and also provides the ability to rearrange packets if they arrive out of order. The time stamp is a 32-bit number (bits 32–63) that reflects the sample interval. If a codec sends one sample per RTP packet, then this value would increment by 1 for each RTP packet. If a codec is used that sends 20 samples at a time, the time stamp field will increment by 20 for each RTP packet sent. The synchronization source field contains a unique 32-bit identifier that is used to identify the current source of the RTP stream. For point-to-point connections in a basic phone call, this value never changes. In a conference bridge where multiple sources can be sending, this value can change over the life of the connection to identify the current sender. The Contributing Source Identifiers is a list of up to 15 sources that could be contributing to the RTP stream. This list is useful in the case of a conference bridge where multiple audio signals are mixed and then sent back to all recipients. This enables all RTP receivers to know all participants currently contributing to the conversation.

As RTP packets traverse an IP network, each packet will be delayed for a time depending upon the traffic conditions in the network. The amount of delay any individual RTP packet will experience can vary significantly. This results in RTP receivers having RTP arrival rates that vary, causing what is known as a *jitter.* A real-time codec needs to have minimal received jitter in order to reconstruct the signal (voice or video) from the far end sender. As a result, RTP framers implement a receive buffer to absorb the jitter of the received RTP stream. Figure 6-11 shows a block diagram of the buffering used by an RTP framer. On the framer's sender side no buffering is necessary,

Bit position	0–1	2	3	4–7	8	9–15	16–31
0	Ver.	P	X	CC	M	PT	Sequence number
32	Time stamp						
64	Synchronization source identifier						
96	Contributing source identifier						
128	. . .						

Figure 6-10 RTP header format

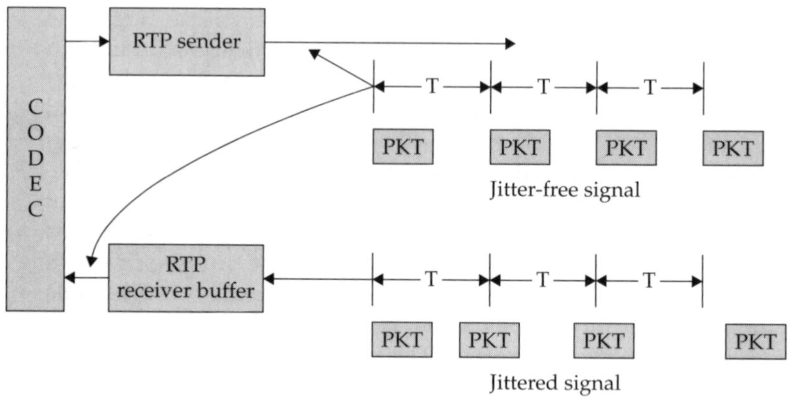

Figure 6-11 RTP framer buffering

since the codec will transmit a signal that is jitter-free. As shown in Figure 6-11, the RTP framer receives a jittered signal, but after buffering, the signal is once again jitter-free. The variation in the arrival of the packets will dictate how much buffering is required to smooth out the received signal. The larger the buffer, the more delay is imposed on the received signal. If the delay gets too large, then that causes noticeable quality problems for interactive applications such as a voice phone call. For example, when the delay approaches a half second, users may start to experience voice echo. Today's IP networks are designed to have minimal end-to-end delay, allowing RTP buffering to be set in the 100 to 200 ms range, which will provide for the proper removal of received jitter without a noticeable delay affecting the quality of service to the users.

RTP normally operates over the UDP protocol because of its low latency and quick establishment procedures. The disadvantage of UDP is that it doesn't provide for congestion control and also has a limited size packet that can be sent. TCP has been defined for RTP in RFC 4571 [22] but is used not nearly as much as UDP for the reasons mentioned. Figure 6-12 shows the protocol stack relationship between various real-time protocols, UDP, TCP, and IP.

6.4.2 RTCP

RTCP (RTP Control Protocol) is a protocol that always goes hand in hand with RTP. In fact RTCP is defined in RFC 3550 along with RTP. RTCP is used to provide quality statistics control information about RTP streams. RTCP packets are sent periodically, using a separate channel from the RTP stream, which contains the latest updated statistics. Usually RTP sends and receives on even port numbers, whereas RTCP uses the next-higher odd port number. These RTCP statistic packets are periodically sent at a much lower rate compared to the rate of the RTP stream. RTCP collects data such as packet counts,

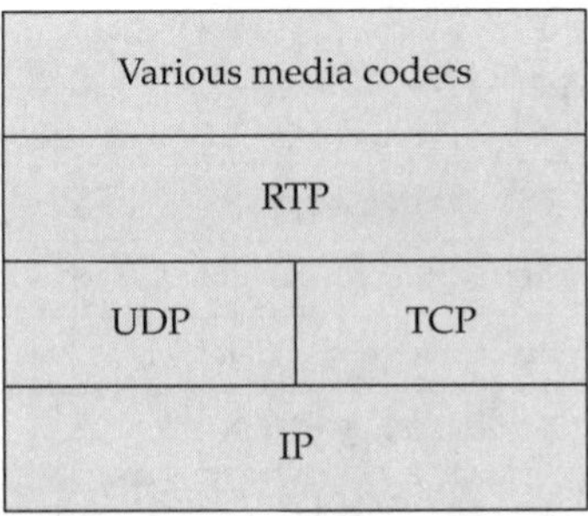

FIGURE 6-12 RTP protocol stack

lost packet counts, jitter, and round-trip delay. Some RTP implementations use this information to make adjustments in QoS parameters of the RTP connection or even to change the codec being used. It is important to note that real-time applications such as voice or video can deal with a certain amount of packet loss without the user noticing; however, the end-to-end delay must be kept within range. Another useful function of RTCP is that it periodically sends the names of the streams associated for each synchronization source, which assists in diagnosing and identifying the source of problem streams. RFC 3550 defines the following types of RTCP packets:

- Sender Reports are sent periodically from RTP participants that are active senders, containing statistics on transmit and receive counters.
- Receiver reports are sent from RTP participants that are listener only and report on their receiver statistics.
- Source description messages are sent from sources identifying their name.
- BYE messages are sent from RTP participants as they drop off of the RTP connection.
- An additional category is used of messages that are application-specific and are defined by the codec's specification.

6.4.3 SRTP/SRCTP

SRTP (Secure RTP) is defined by RFC 3711 [23] as an RTP profile that offers a higher level of security using encryption and authentication. Included in the same RFC is SRTCP (Secure RTCP) to increase the security for the RTCP status messages. SRTP allows for flexibility whereby most of its attributes relating to encryption and authorization can be enabled or disabled. However, message authentication is always required.

Figure 6-13 shows the SRTP frame format from RFC 3711. Most of the frame format is the same as the format for the base RTP protocol

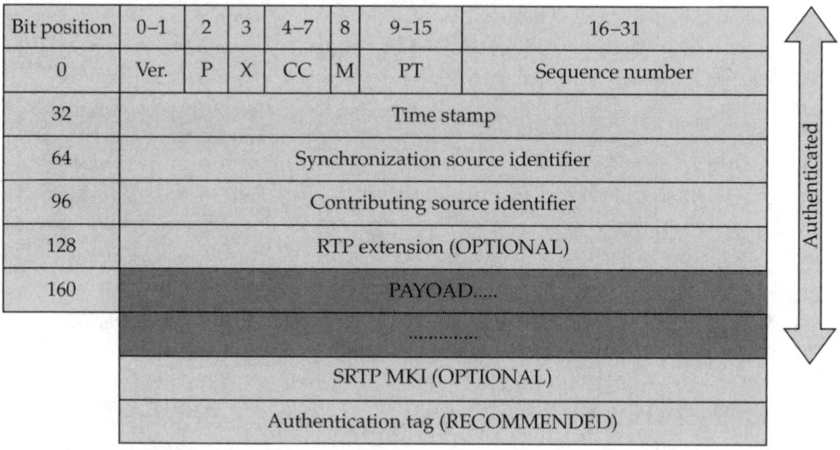

Bit position	0–1	2	3	4–7	8	9–15	16–31
0	Ver.	P	X	CC	M	PT	Sequence number
32	Time stamp						
64	Synchronization source identifier						
96	Contributing source identifier						
128	RTP extension (OPTIONAL)						
160	PAYOAD.....						
	....						
	SRTP MKI (OPTIONAL)						
	Authentication tag (RECOMMENDED)						

Authenticated

Figure 6-13 SRTP frame format

as shown in Figure 6-10. The following are the added fields of the SRTP profile:

- Master Key Identifier (MKI) identifies the master key used for encrypting. This field is optional, based on the configuration usage of SRTP.
- The authentication tag has a configurable length and is highly recommended. This tag is used to carry message authentication data. Figure 6-13 shows that the authentication applies to all fields except for the MKI and authentication tag fields.

The dark-colored portion of Figure 6-13 is the encrypted section of the message, which applies to the RTP payload contents.

References

[1] ITU-T G.711 "Pulse Code Modulation (PCM) of Voice Frequencies".
[2] ITU-T Recommendation G.718 "Frame Error Robust Narrowband and Wideband Embedded Variable Bit-Rate Coding of Speech and Audio from 8-32 Kbit/s", May 2008.
[3] ITU-T Recommendation G.722 "7 kHz Audio–Coding within 64 kbit/s", 1993.
[4] ITU-T Recommendation G.729 "Coding of dpeech at 8 kbit/s Using Conjugate-Structure Algebraic-Code-Excited Linear Prediction (CS-ACELP)", July 2009.
[5] ITU-T Recommendation G.721 "Adaptive Differential Pulse Code Modulation (ADPCM) for Audio Encoding", 1984.
[6] ITU-T Recommendation H.261 "H.261 : Video Codec for Audiovisual Services at px64 kbit/s", March 1993.
[7] ITU-T Recommendation H.120 "H.120 : Codecs for Videoconferencing Using Primary Digital Group Transmission", March 1993.
[8] ITU-T Recommendation H.262 "H.262 : Information Technology-Generic Coding of Moving Pictures and Associated Audio Information: Video", February 2000.
[9] ITU-T Recommendation H.263 "H.263 : Video Coding for Low Bit Rate Communication", January 2005.

[10] ITU-T Recommendation H.264 "H.264 : Advanced Video Coding for Generic Audiovisual Services", March 2010.
[11] IETF RFC 2778 " A Model for Presence and Instant Messaging", February 2000.
[12] IETF RFC 3860 " Common Profile for Instant Messaging (CPIM)", August 2004.
[13] IETF RFC 3862 "Common Presence and Instant Messaging (CPIM): Message Format", August 2004.
[14] IETF RFC 2368 "The mail to URL scheme", July 1998.
[15] IETF RFC 3428 "SIP Extension for Instant Messaging", December 2002.
[16] IETF RFC 4975 "The Message Session Relay Protocol (MSRP)", September 2007.
[17] IETF RFC 2046 "MIME Part Two: Media Types", November 1996.
[18] IETF RFC 4103 "RTP Payload for Text Conversation", June 2005.
[19] ITU-T T.140 "Presentation Protocol for Text Conversation Application", 1998.
[20] IETF RFC 1889 "RTP: A Transport Protocol for Real-Time Applications", January 1996.
[21] IETF RFC 3550 "RTP: A Transport Protocol for Real-Time Applications", July 2003.
[22] IETF RFC 4571 "Framing RTP and RTCP Packets over Connection–Oriented Transport", July 2006.
[23] IETF RFC 3711 "The Secure Real-Time Transport Protocol (SRTP)", March 2004.

CHAPTER 7

Femtocell Security Solutions

Femtocell networks utilize wireless and fixed broadband access technologies to offer their unique benefits. Both networks by themselves conventionally have security issues that must be addressed. As a result, femtocell networks are vulnerable to security concerns that occur in both wireless and fixed broadband networks. This chapter will describe the main security vulnerabilities of femtocells and describe some of the common technologies used to deal with them.

7.1 Femtocell Security Vulnerabilities

There are several general aspects of security that all networks need to be concerned about. Some of the main aspects are

- **Voice and data confidentiality** All telecommunication networks need to protect the privacy of the user. For voice calls, the conversation between the parties should not be intercepted and heard from any third party. The exception to this is a legal wire tap. For data sessions such as a user surfing the web, the information exchanged should be readable by only the user and not obtained by a third party for malicious intent.
- **Content integrity** Any information sent from a user's device should not have its content altered by another third-party device that is unintended. An example of this would be maliciously changing a user's caller ID for an outgoing call where the called party is misinformed as to who the calling party is.
- **Theft of service** No third party should be able to steal a user's identity from the network. Once a user's identity has been stolen, then the third party can use the services offered by the network on the behalf of the user free of any charges. The act of emulating another user is called *spoofing*, which could result in very expensive theft of service.

- **Service availability** One of the key quality metrics customers use is service availability. *Service availability* is a benchmark based on the percentage of time the network is fully capable of offering service. Denial of service occurs when a protocol attack is invoked on a key network element that causes service disruption to users. The attack usually is in the form of overloading the network element with large amounts of protocol traffic to the point that the network device can no longer function properly.

In order for a network security breach to occur, the intruder must gain access to the network. Figure 7-1 shows some of the vulnerable areas where an intruder can directly or indirectly attempt to breach security of a femtocell customer. Shown in the figure are the symbols *T, D, S,* and *I* that indicate the likely type of attack that can be attempted. *T* is for theft of service, *D* is for denial of service, *S* is for snooping attempts, and *I* is for intercepting and modifying a user's content.

The most obvious access point for an intruder is over the air interface, since the accessibility is simply to be within the radio range of the femtocell. Common threats over the air interface are the theft of service and snooping. Air interface intrusions are avoided by using ciphers to

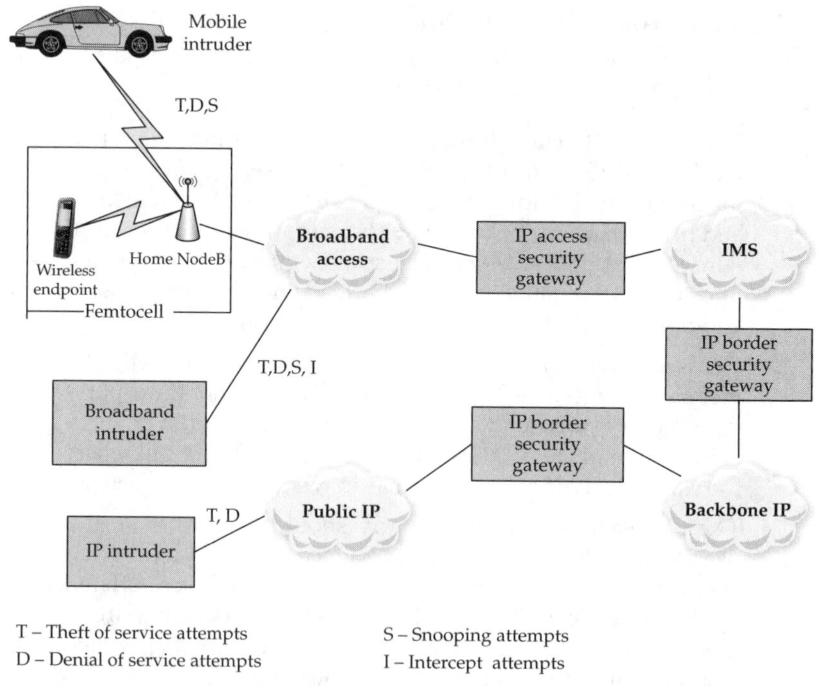

Figure 7-1 Femtocell security attempts

encrypt the signals transmitted between a wireless endpoint and the femtocell Home NodeB. The section "Air Interface Ciphers" of this chapter provides details on different types of ciphers and their implementation.

Another access point for an intruder is at the broadband access interface. Some broadband access technologies, such as cable, use a shared access interface that could allow intruders to attempt theft of service or to snoop a customer's traffic. Another easy access point for a potential intruder is through the public Internet. A femtocell Home NodeB uses IPsec to encrypt communications between itself and an IP access security gateway, allowing it protected access to the IMS network. The sections of this chapter dealing with IPsec provide details on different implementation options that exist for IPsec.

All of the access points mentioned provide opportunities for denial of service (DoS) attacks by a potential intruder by simply generating a large amount of traffic directed at any one device, especially a Home NodeB. To protect against DoS attacks, security gateway devices at both the access network and the IP borders in the core would implement DoS protection. DoS protection works by monitoring the rate of traffic from all sources. If any one source becomes overactive, the first security gateway that receives traffic from that source will treat it as misbehaving and drop all of its traffic. This DoS protection shields the rest of the network from the DoS attack, avoiding service disruption. More details on traffic rate monitoring can be found in the section "Traffic Controls" in Chapter 8.

7.2 IPsec

IPsec is a set of protocols that provides for IP packet security in the areas of both integrity and confidentiality. One of the complexities with the set of IPsec protocols is the numerous tunable parameters along with the various protocol stack combinations. IPsec is run between two networking devices that require a higher level of security when the communications between them are over an unsecure IP network. The following sections describe some of the common IPsec protocol stack uses.

7.2.1 Authentication Header (AH)

The authentication header (AH), as the name implies, offers authentication of IP packets. The AH does not provide for confidentiality, since it doesn't encrypt the packet in any way. Authentication affords a recipient of an IP packet the ability to verify if the packet has been tampered in transit from the sender. If a packet's contents has been modified in the network, thereby compromising the integrity of the sender's data, the IPsec AH will be able to detect the change and the receiver will discard the packet so that the application doesn't process invalid data. The AH is able to detect that packet contents have been tampered with by generating a cryptographic hash authentication code based on the packet contents and including this generated code in the IP packet's AH header. When a AH packet is received, the hash code is generated at the receiver's end based on the contents of the packet. If the hash code generated by the sender doesn't match the hash code regenerated by the receiver, the packet has been tampered with and is discarded. If both codes match, then the received packet has been properly validated and can then be processed without concern of breach of integrity.

The AH can be used in two different modes, known as transport mode and tunnel mode. Transport mode provides for authentication by simply inserting an AH header in the IP packet between the IP header and the IP packet payload. Figure 7-2 shows an example of a TCP packet before and after the packet has an AH header inserted. The AH header components, which appear at the center right, include the following fields:

- The Next Header field specifies the protocol of the payload being used, which is obtained from the IP packet header Protocol field prior to the AH header being inserted. The IP packet header Protocol field is modified to specify the type of IPsec mode that is being used.

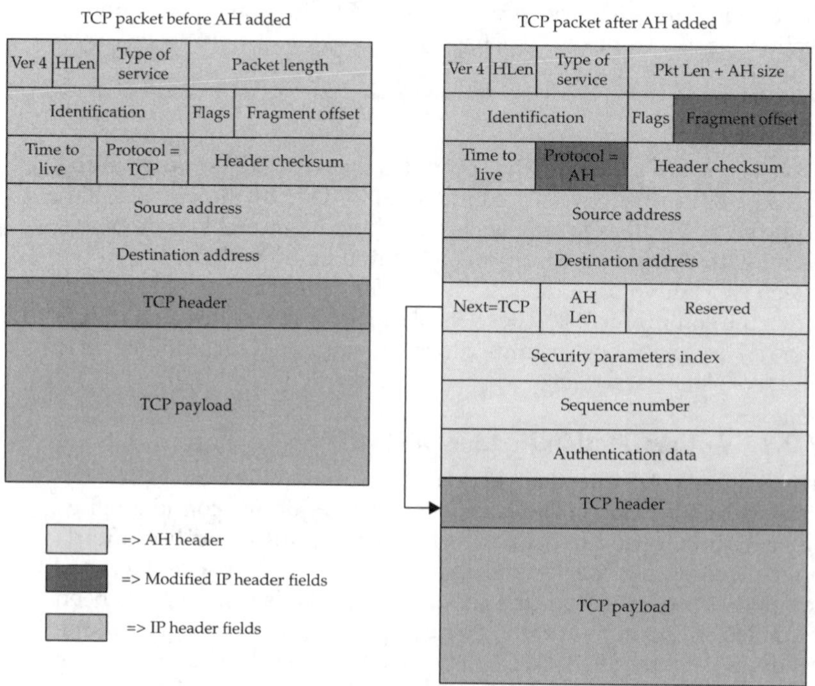

FIGURE 7-2 AH transport mode for a TCP packet

- The AH Length field specifies how many 32-bit words are in the AH header.
- The Reserved field is set to zero and is an extra space in the AH for future use.
- The Security Parameters Index is a 32-bit number that identifies the IPsec association between a source and a destination. More details are provided in the section "IPsec Associations."
- The sequence number is a number that increases by 1 for each AH packet sent for an IPsec association. This helps in identifying if any changes have occurred.
- The Authentication Data field holds the generated hash value populated at the sender and recomputed at the receiver to validate the integrity of the packet. Included in the hash calculation is the IP packet payload; the AH header except for the authentication data field; and most of the IP header, except for the TOS, the flags fragmentation offset, the TTL, and the IP header checksum.

The IP header, as shown in Figure 7-2, has two fields that are modified for the transport mode, which are the Protocol and Packet Length fields. The Protocol field value is moved to the next field of the AH header, and the protocol is changed to the type of IPsec being used; in this case, it is set to AH.

The other mode supported by AH is called tunnel mode. In tunnel mode the complete original IP packet is carried as a payload after the AH header is inserted after the IP header. Figure 7-3 shows an example of a TCP packet before and after the packet has an AH header inserted using tunnel mode. The IP header's protocol is also modified to AH to indicate that AH is being used. The Source and Destination Addresses of the IP header are changed to the source and destination addresses of the IP tunnel. A tunnel is formed by changing the IP header's address such that all tunnel packets are sent between two tunnel gateway devices. The original IP packet is piggybacked onto the modified AH tunnel packet. Tunneling is widely used in VPNs (virtual private networks) because it allows two private networks to be securely joined over an unsecure IP network.

In the AH tunnel mode the authentication hash data is generated on the tunnel origination side. The same fields as the transport mode are used in the hash function calculation as well as the additional IP header fields after the AH header. At the destination tunnel end a regenerated hash calculation is performed and its value is compared to the Authentication Data Value. If the values don't match, the packet is discarded. If the hash values match, the tunnel destination device strips off the tunnel IP header and the AH header, leaving the original packet, which can now be routed as if the two ends of the tunnel were directly connected.

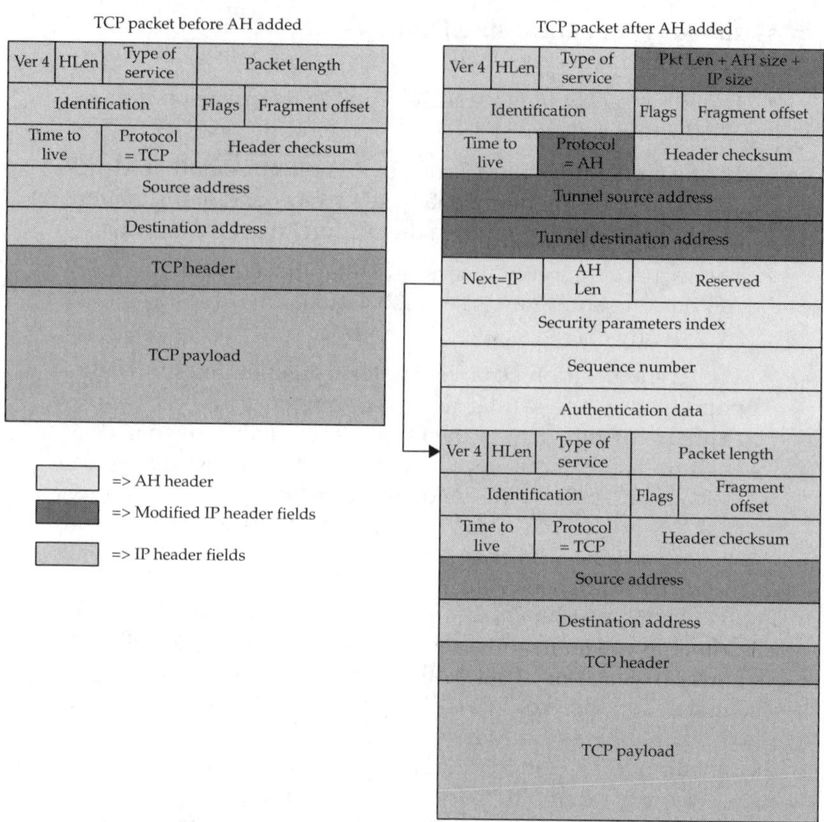

FIGURE 7-3 AH tunnel mode for a TCP packet

The Next field in the AH header distinguishes what type of AH payload is being carried. In tunnel mode the next field is always set to IP, indicating that a complete IP packet is being carried. In the case of transport mode, the next field will carry the value that was in the IP protocol field of the original IP packet before AH packet modifications. In the example of Figure 7-3, the next field is set to TCP, since the packet payload data transported is TCP.

7.2.2 Encapsulating Security Payload (ESP)

ESP provides for confidentiality using encryption of the packet payload. Encrypting the packet payload prevents eavesdropping when packets traverse an unsecure IP network. Confidentiality is critical for applications where part or all of the information exchanged contains private information not intended to be shared. The encryption algorithm is one that is configured as part of the IPsec association. There are many types of encryption algorithms that are described in the section "Air Interface Ciphers" in this chapter. ESP also offers an optional authentication through the same hash generation approach as AH. ESP also supports both a transport mode and a tunnel mode.

Figure 7-4 shows an example of a TCP packet using ESP transport mode. In the IP header the Packet Length and the Protocol fields are modified for ESP. The Protocol field is changed to ESP. Similar to AH, ESP inserts a Security Parameters Index and a Sequence Number just after the IP header. If authentication is being used, the hashed authentication data field is appended to the end of the packet. The original IP payload, in this case including the TCP header and TCP payload, is encrypted as shown in the lower right in Figure 7-4. After the payload is encrypted, padding of the payload is necessary, since the encryption algorithms result in block size output, which is usually longer than the original payload. All other fields are not encrypted, allowing for easy address translation if necessary in places where NAT (Network Address Translation) is performed.

Figure 7-5 shows an example of a TCP packet using ESP tunnel mode. The differences between transport mode and tunnel mode are the same as for AH. In the IP header the Packet Length and the Protocol fields are modified for ESP. The Protocol field is changed to ESP. The Source and Destination Addresses of the IP header are

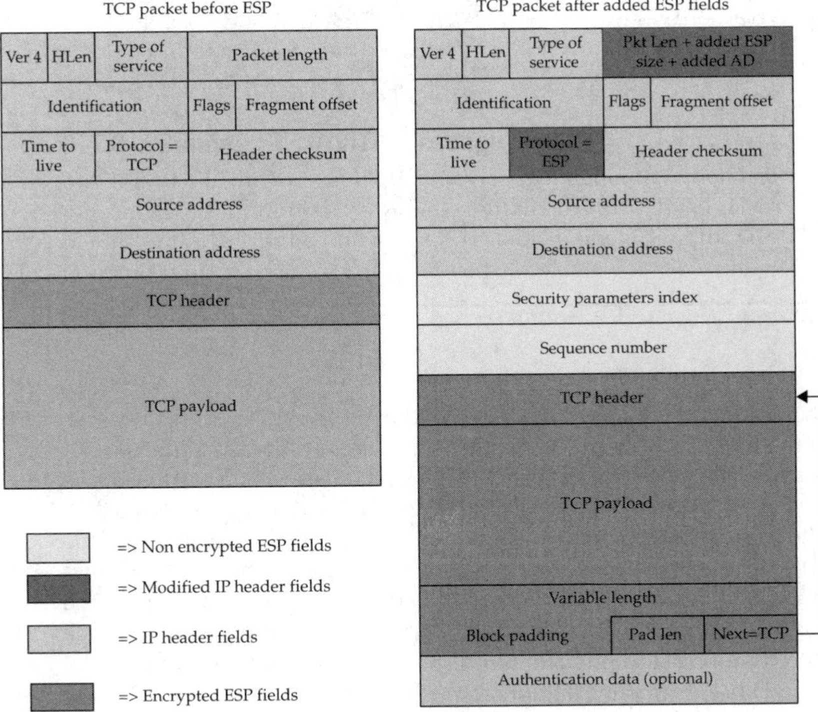

FIGURE 7-4 ESP transport mode for a TCP packet

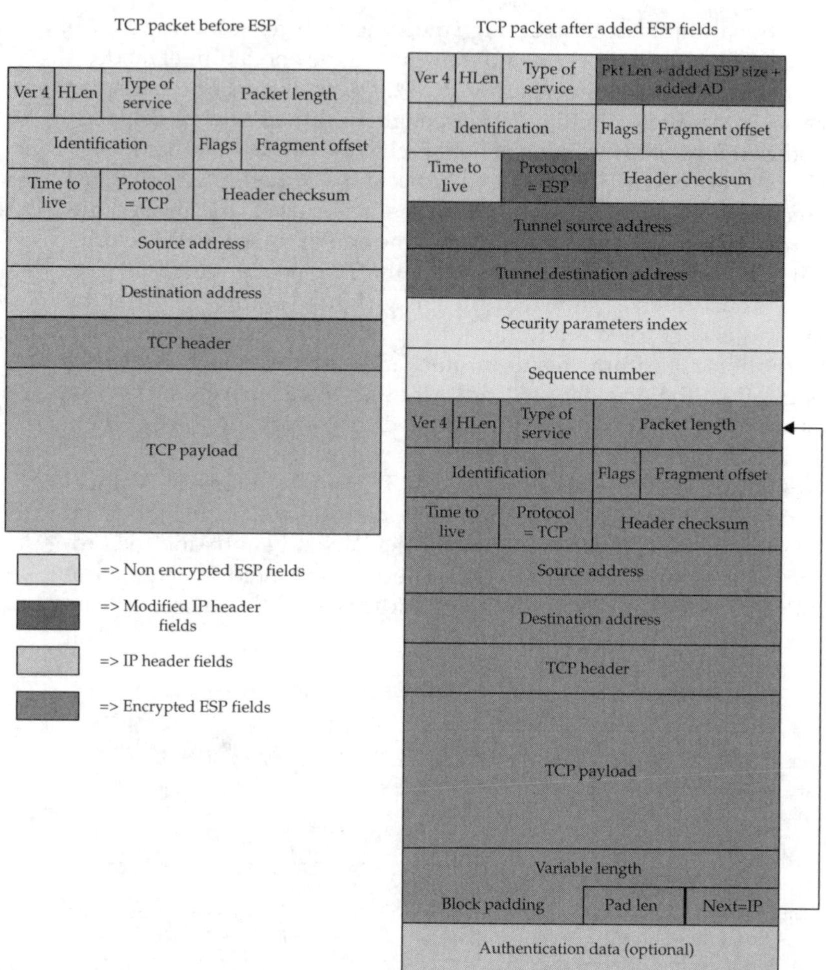

Figure 7-5 ESP tunnel mode for a TCP packet

changed to the source and destination addresses of the IP tunnel. Similar to AH, ESP inserts a Security Parameters Index and Sequence Number just after the IP header. If authentication is being used, the hashed authentication data field is appended to the end of the packet. The original full IP packet is carried as an encrypted payload in the tunnel packet. The encrypted part of the packet is shown in the lower right in Figure 7-5. After the payload is encrypted, padding of the payload is necessary, since the encryption algorithms result in block size output, which is usually longer than the original payload. All other fields are not encrypted, allowing for easy address translation if necessary.

7.2.3 IPsec Associations

An IPsec association is a unidirectional flow of agreed-upon encryption algorithm and associated configuration options. A typical two-way communication path would require two IPsec associations. Normally, both directions would be negotiated and configured with the same configuration parameters. IPsec associations can be identified by the source and destination IP addresses along with the Security Parameters Index (SPI) of each of the two devices that are engaged in a secure IPsec channel. The SPI is especially important when two devices have more than one IPsec association between them, where the SPI uniquely identifies each of the associations. When an encrypted message is received, a lookup is performed on the sending IP address and the SPI, allowing the receiver to set the decryption parameters.

7.3 Cryptographic Hash Algorithms

A cryptographic hash algorithm is used to take a variable-length message that contains arbitrary data as an input and produce a fixed-size hash value that is unique. Figure 7-6 shows a high-level picture of a hash function in use with a variable-length message, resulting in a fixed-size output called a *digest.*

Hash algorithms have several uses in terms of networking security, especially in the areas of authentication. As seen in the section "IPsec," the Authentication field output is the result of a hash function computation. Figure 7-7 shows a block diagram of a hash algorithm. At the input is a variable-length message that is first broken into fixed-size blocks. The last block is padded so that it is the same length as all the other blocks. The reason the message is broken into fixed-size blocks is that the hash function operation processes a fixed number of bits at a time.

The algorithm shown in the diagram starts on the left with the initial values for a key along with the first fixed-size block used in the first hash function computation. The output of the first hash function is then used as input to the second hash function along with data for the second fixed block. This continues until the last block, which in this example is the fifth fixed-size block with padding. Each hash function is a series of logical (ORs, Exclusive ORs, Anding) operations and shifting of bits. The end result is a unique fixed-size value called the digest that can be used to uniquely identify the contents of the message.

FIGURE 7-6 Hash function overview

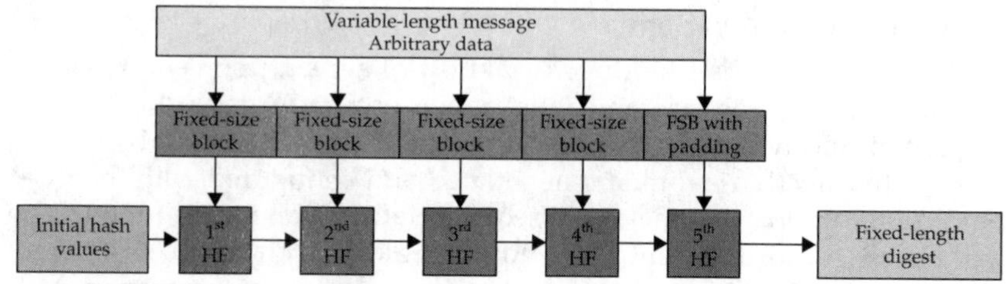

FIGURE 7-7 Hash algorithm block diagram

MD5 is one of the most widely used cryptographic hash functions that results in a 128-bit hash value. MD5 is specified in RFC 1321 [1] and is used in many security applications, including file integrity checks and IPsec authentication. Since the time MD5 was designed, weaknesses have been discovered such as the number of duplicate hash values for different value inputs. This weakness limits the usage of MD5. Other hash algorithms such as the SHA [2] family of hash functions are now being recommended.

There exists a set of SHA hash functions that are called SHA-0, SHA-1, and SHA-2. The SHA algorithms were designed by the National Security Agency (NSA) and published by NIST (the National Institute of Standards and Technology). The SHA-2 algorithm uses a variable-size digest in order to minimize the duplication of hash values. NSA will also be coming out with a SHA-3, which is intended to have improved security qualities. Table 7-1 is a list of cryptographic hash algorithms showing the digest size, the fixed block size, and whether the algorithm is collision resistant or not. An algorithm is collision resistant if two different messages cannot be found that produce the same digest result. Since hash algorithm digest values

Algorithm	Digest Size	Fixed Block Size	Collision Resistant
MD2	128	128	No
MD4	128	512	No
MD5	128	512	No
RIPEMD	128	512	No
RIPEMD-128/256	128/256	512	Yes
RIPEMD-160/320	128/256	512	Yes
SHA-0/1	160	512	No
SHA-256	256	512	Yes
SHA-512	512	1024	Yes

TABLE 7-1 Hash Algorithms

are intended to verify that message contents uniquely match, having collision resistance is a very important quality.

7.4 Air Interface Ciphers

In wireless networks the most obvious place to breach communications is at the air interface, since it is easiest to access. As a result, lots of encryption technology has been developed in this area. *Encryption* is a process where information, including plain text, will be scrambled before it is transmitted so that anyone attempting to eavesdrop cannot interpret the sender's message. On the receiver's side the scrambled message is unencrypted by performing the reverse procedure. The most common encryption technology is block ciphers that encrypt the air interface communication between a wireless terminal and the base station or its equivalent in the case of a Home NodeB. This section will describe the basics of block ciphers and provide information on some of the commonly used ciphers.

Figure 7-8 shows a diagram of a block cipher. On the sender's end messages that contain either voice or data are encrypted using a key. The encrypted signal is then safely sent over a communication channel, which can be an air interface. On the receiver side the signal is decrypted using the same key as the sender's, resulting in the original voice or data message being reproduced. The key that is used is a secret key known only by the two devices communicating. If the secret key is somehow discovered by a third party, then it is possible for the third party to decipher the communication stream. In order to minimize the chances of a key being identified, it is common practice to regenerate and exchange a new key any time two devices enter into a secure communications channel. An important part is the exchange of a key between two entities in a manner that the key itself cannot be

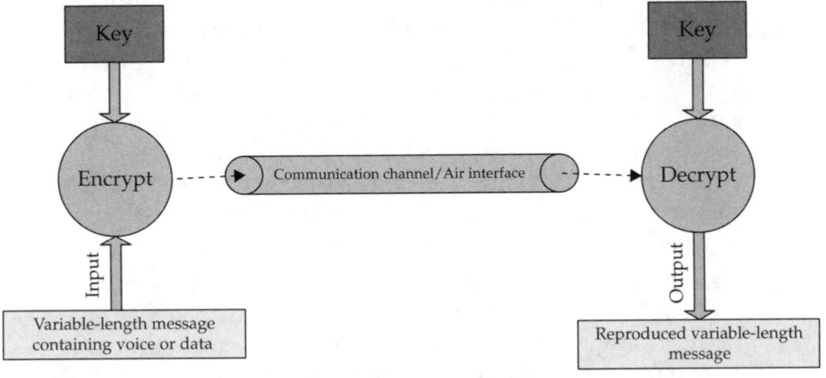

FIGURE 7-8 Block cipher block diagram

intercepted, since an encrypted channel has not been established yet. The section "IKE" in this chapter describes one commonly used approach for key exchange and how key interception is avoided.

7.4.1 Block Ciphers

A *block* cipher encrypts a block of data, such as 128 bits, at a time. Another very commonly used category of ciphers is a stream cipher. The following briefly describes some of the more common block ciphers used in wireless communications today.

- DES is an acronym for Data Encryption Standard, which is one of the first standardized block ciphers that was developed by IBM. The DES standard was published by the NIST (National Institute of Standards and Technology), formerly known as NBS (National Bureau of Standards) in the form of a FIPS (Federal Information Processing Standard) [3]. DES uses a 56-bit key, which by today's standards is quite small. Since DES uses a small key, it has been proven to be cracked in less than a day. Improvements have been made to DES to increase its key size; as a result, triple DES is far more secure. Triple DES is sometimes used in WiMAX applications.
- A block cipher modeled after DES is the AES (Advanced Encryption Standard) [4]. AES has a family of ciphers called AES-128, AES-192, and AES-256. They all use 128-bit block sizes; however, they have variable key sizes of 128, 192, and 256 bits, as implied by their names. AES block ciphers are widely used, including in 3G, LTE, and WiMAX air interfaces. AES was also standardized and published by NIST and is approved for use by the NSA for the transport of top-secret information. It took over five years to standardize DES from a selection process of the best encryption algorithms from around the world.
- Blowfish is a block cipher that was designed during the same time frame as AES and also addresses the weaknesses of DES. It is a public-domain cipher that doesn't require any royalties for its use. One of the benefits of Blowfish is its encryption rate, which is much faster than DES. The basic operations of the algorithm use XOR, ADD, and MOV operations that can be performed in today's processor caches for faster performance. Blowfish uses a 64-bit block size and supports a varying key length up to 448 bits.
- Kasumi, also known as A5/3, is used in GSM for GPRS and also for 3GPP wireless networks. Kasumi was developed by SAGE (Security Algorithms Group of Experts) under the auspices of ETSI (European Telecommunications Standards Institute) based on an existing algorithm called MISTY. The name Kasumi was

chosen because it translates into mist in Japanese. Kasumi uses a 64-bit block size and a key of 128 bits.

7.4.2 Stream Cipher

A *stream* cipher encrypts one bit at a time, by contrast with a block cipher that encrypts a large number of bits at a time. Stream ciphers tend to execute at higher speeds because of less complex hardware to implement the algorithm. However, the cost of a simpler design results in stream ciphers being more susceptible to security breaches. A5/1 and A5/2 use a 54-bit key, with A5/2 having improved significantly on the computational speed. Both are used in GSM wireless networks used in the United States and Europe. The A5/1 algorithm uses a set of three linear feedback shift registers (LFSR) with a technique known as irregular clocking of the shift registers. There have been published cases of A5/1 encoding being cracked, resulting in attackers being able to listen in on GSM mobile phone conversations.

7.5 Cryptographic Legal Issues

One of the outcomes of World War II is that cryptography is regulated by many countries, including the United States, because of the national security issues raised by the ability to decipher all means of communication. Prior to the wide use of the Internet this policy was of no consequence to the average person. The Internet browsers today utilize fairly advanced cryptography, including transport layer security (TLS), secure socket layer (SSL), and secure/multipurpose internet mail extensions (S/MIME), which in the past would have been violations of the cryptography regulations. Because of the Internet's wide use, the regulations on cryptography have been significantly relaxed. Still, today export of cryptography technology needs to be registered for many governments. For some governments, uses of cryptography, including secured Internet access, are very tightly restricted, and in some cases, they are prohibited.

7.6 IKE

IKE (Internet Key Exchange), including IKE version 2, is a protocol used to exchange private keys for the purpose of establishing an IPsec association. IPsec uses a cipher to encrypt its packet contents to prevent any third party attempting to intercept and interpret the private communications between the two terminations of the IPsec association. Both terminations of the IPsec association need to know the secret key in order to successfully communicate. This is called a symmetrical key algorithm, since both the encrypting side and the decrypting side use the same secret key. The secret key can be provisioned ahead of time for associations that are intended to be static and are expected to live for a long time. For IPsec associations that are short-lived and occur frequently, preprovisioning the secret key is not practical. This is where IKE is used to exchange private keys that are based on the Diffie-Hellman key exchange [5]. A public key is first used to encrypt messages that carry private keys to be exchanged.

Public-key cryptography uses the concept of an asymmetrical key algorithm where a sender encrypts a message using a recipient's public key to encrypt a message. When the recipient receives the message it uses its private key to decrypt it. Using public-key cryptography, only the recipient's private key can decrypt the message. As long as the private key of the recipient is kept confidential, then messages can be safely sent without worries of a third party being able to decrypt the contents of the message. Public-key algorithms are widely used on the Internet, including SSL and TLS. Figure 7-9 shows a high-level diagram of asymmetric key generation. A large random number usually is used as input to an asymmetric key generator. As shown in the figure, the output of the key generator is a public key and a private key. The public key is used by any sender wishing to encrypt a message and send it to the owner of the private confidential key. When the encrypted message is received, it can be decrypted only by the private key, thereby keeping the communications confidential.

IKE version 2 uses as few as three pairs of messages to establish a secured association between two entities. Figure 7-10 shows this minimum message exchange and the following describes each message.

1. IKE_SA_INIT is an initiation message to start an IKE Security Association. The message contains the IKE Header (HDR), the Security Association (SA), the Key Exchange Information (KE), and the Initiators nonce (Ni), which is a unique random number. The header contains the Security Parameters Index (SPI).

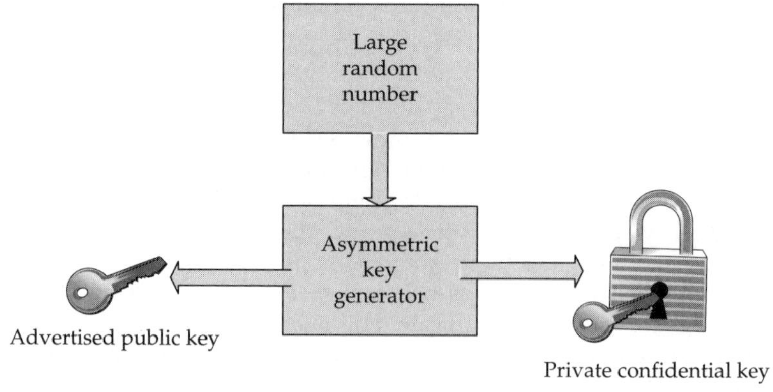

Figure 7-9 Asymmetric key generation

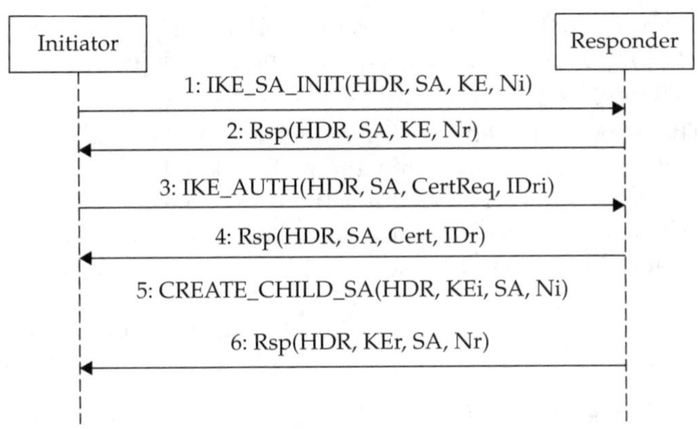

FIGURE 7-10 IKE security association sequence

2. The Responder processes the IKE_SA_INIT by choosing a cryptographic scheme from the initiator's list of choices and sends back a response message. The response message contains the IKE header (HDR), the Diffie-Hellman key exchange (KE), and the response nonce. At this point the keys have been exchanged and the cryptographic parameters have been negotiated, allowing each side to begin encrypting content. All the messages that follow after this message are encrypted except for the header and are integrity checked as well.
3. The initiator sends an IKE_AUTH message intended to authenticate the first pair of messages, exchange identifiers, and if necessary, exchange certificates. Since the encryption keys were exchanged in the first pair of messages, the confidential parts of this message are encrypted. The initiator sends its identity in the encrypted payload as IDri and sends a CERTReq if requesting certificates. This message can also be used to send further encrypted information that is useful in configuring both ends of the secured IP peering association.
4. After the responder processes the IKE_AUTH message, it sends a response message back. The response message contains the IKE header, the security association, a certificate if requested, and the responder's identification as IDr. At this point both sides have exchanged their private keys successfully and have authenticated with each other. These private keys can now be used to encrypt IPsec sessions between the two peers as described in the section "IPsec".
5. If at any point either peer of the Security Association deems it is necessary to exchange another pair of private security keys for the purpose of added security, then child security associations

can be created. This is done by either side sending a CREATE_CHILD_SA message that requests another Diffie-Hellman by including the private key KEi and the nonce Ni.

6. The responder processes the requested SA child request and replies with the newly generated child key Kr along with a response nonce. At this point the child security association private key pairs can be used to establish further IPsec associations between the two IP peers.

7.7 UICC

The UICC (Universal Integrated Circuit Card) is a small circuit card hosting a device called a smart card that is used in wireless terminals including GSM and UMTS. UICC is capable of storing personal and confidential information to assist in offering wireless services. GSM wireless devices use a UICC to contain a SIM (Subscriber Identity Module), and UMTS networks use it to host a USIM (Subscriber Identity Module). IMS terminals require the UICC to host an ISIM (IMS Subscriber Identity Module), and CDMA networks use a CSIM (CDMA Subscriber Identity Module). The UICC allows users to change phones by simply swapping cards.

One of the key pieces of information stored in a SIM/USIM is an IMSI (International Mobile Subscriber Identity), which is a unique number that identifies a mobile user. The IMSI is used during registration with the Home Location Register. Security Authentication key information as well as ciphering data is stored on the UICC. This security information is used by the EAP (Extensible Authentication Protocol), which is an authentication framework implemented in wireless networks. GSM uses EAP-SIM (EAP for GSM Subscriber Identity) for authentication and key distribution. UMTS uses EAP-AKA (EAP for UMTS Authentication and Key Agreement) for the same purpose. Both EAP-AKA and EAP-SIM are based on IKE for key distributions, allowing IPsec tunnels to be established to ensure secured communications.

Using EAP, femtocells have a proven approach to authenticating wireless devices as they enter into a femtocell. A similar approach can be used for registering Home NodeB femtocell devices. EAP has been proven to be reliable and allows wireless service providers the ability to leverage an existing technology. Home NodeBs can be also designed with flash memory instead of UICC smart cards to store identification and secret keys for authentication to be used during registration as well. This type of data must be securely maintained.

7.8 Home Device Security

Since femtocell devices are physically accessible by potential security hackers, they need to be as tamperproof as possible. Intelligent hackers will attempt to reverse-engineer home femtocell devices to control access to a network for various reasons, including theft of service and breach of confidentiality. Would-be hackers can attempt to load illegal software to alter the behavior of the device. Other attack approaches are to obtain or modify authentication data, allowing a hacker to emulate another user or obtain private customer information that is stored on the home device. Home femtocell devices need to be carefully designed to minimize hackers' chances of intrusion. Most design approaches should include authentication against encrypted passwords or access codes. Areas that can be authenticated include access to all management interfaces to the device, any software that is loaded onto the device, and access or modifications to secured keys or identification IDs.

References

[1] IETF RFC 1321 "The MD5 Message-Digest Algorithm", April 1992.
[2] FIPS Publication 180-1 "Secure Hash Standard", April 1995.
[3] FIPS Publication 46-3 "Data Encryption Standard", October 1999.
[4] FIPS Publication 197 "Advanced Encryption Standard", November 2001.
[5] IETF RFC 2631 "Diffie-Hellman Key Agreement Method", June 1999.

CHAPTER 8

Quality of Service

In today's wireless and IP networks many different types of media streams are used by a wide range of applications. The end-to-end network performance characteristics can drastically impact users' experiences of these applications. Some network performance characteristics are more important than others, depending on the application in use. The following are some of the key performance metrics that can impact the users' experience and are typically used to measure the Quality of Service (QoS).

- **Delay** The end-to-end delay, typically measured in milliseconds, is an average value of how long a packet will take to be delivered across the network. Interactive applications like voice and video conferencing will have very poor user experiences if the delay gets too large.
- **Packet jitter** As packets traverse the network, significant variance in the delays can occur. This network delay variation causes packet arrival times to vary at the destination in what is known as packet jitter. With significant packet jitter, applications are forced to delay streaming or drop late packets, which can impact the user experience.
- **Packet loss** In wireless or IP networks packets can be dropped as a result of network congestion or as a result of transmission errors. Some applications are more sensitive to packet loss compared to others. For example, a file transfer would require the packet to be retransmitted, whereas a single voice codec packet being lost would for the most part go unnoticed unless there are bursts of lost packets.
- **Sequential packet order** In IP networks packets can be routed differently from one packet to the next. Normally, consecutive packets will take the same path. At times routes are changed due to network congestion or some failure occurring in the network. When packets take different routes from source to destination, different network delays can result for each of the routes. This difference in delay at times results in packets arriving at the destination out of order. This requires applications to deal

with this problem by either delaying the processing of packets until all are received or dropping the packet that was received out of order. Both solutions can impact the quality of service and the users' experience of the application.

Today IP is being used to carry all types of traffic, including voice, video, and data applications. When IP was originally designed, it was intended for best-effort traffic, which was sufficient for data traffic. For real-time interactive applications like voice and video conferencing, the end-to-end delay and jitter need to be minimal. In order to offer some streams better networking performance, one or more of the QoS mechanisms described in this chapter need to be implemented in the IP network.

8.1 Resource Reservations

Most of today's network transmission technologies are quite resilient and experience minimum amount of transmission errors. A good example of this is fiber-optic networks, which have bit error rates less than one error in a trillion bits sent. When packets are dropped or significantly delayed in an IP packet network, this is typically a result of some form of network congestion that has occurred. The types of congestion that can occur are known as node or link congestion. *Node* congestion means a networking device, such as a router or an IMS I-CSCF, is overloaded with too much traffic to the point it needs to drop or significantly delay packets in order to continue processing packets as fast as possible. *Link* overload is a condition where too many packets are routed to a link between two nodes, where either the bandwidth of the link is exceeded or the network interface capabilities of the equipment on either side of the link are exceeded. In both forms of congestion, Quality of Service will degrade as a result of packets being dropped or delayed.

One mechanism in preventing network congestion that in turn avoids Quality of Service degradation is to reserve resources for application media streams. By keeping track of inventory of all media streams within a packet network, as well as knowing the link and node limitations of all equipment that makes up a network, most if not all congestion can be avoided by reserving resources as needed. Using a reservation approach, any new media stream being requested will only be added to the network provided a path can be found through the network where the resources needed to honor the stream are available.

8.1.1 Policy Servers

Policy Servers are used to keep track of the network resources. Any time a media stream is needed, a request is placed to the Policy Server that contains a complete description of the type of stream that is needed. Included in the request are the bandwidth requirements,

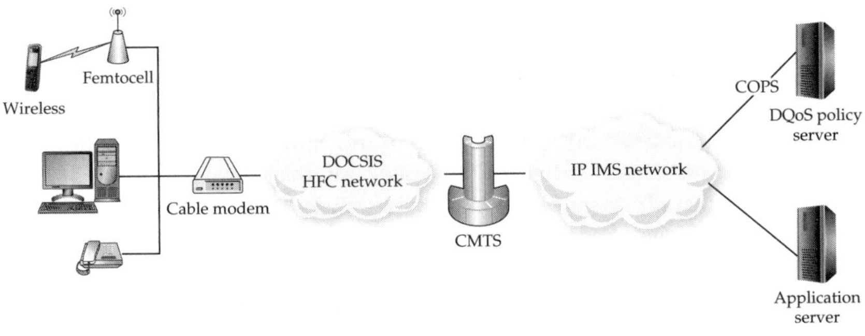

FIGURE 8-1 DQoS Policy Server

the type of traffic such as constant or variable bit rate, and possibly the delay and jitter requirements. As streams are added and deleted the Policy Server keeps track of network resources allocated. The Policy Server can reject a stream request if network resources have been depleted to the point where adding the requested stream could cause network congestion. Using a Policy Server to track network resources allows it to make intelligent decisions for when to admit a new stream or not, thereby avoiding potential congestion conditions.

Policy Servers are used in many types of networks, including cable networks, and their use is specified in the DOCSIS (Data over Cable Service Interface Specification) [1]. Figure 8-1 shows a DQoS (Dynamic Quality of Service) Policy Server used by a cable access network to manage bandwidth allocation on a HFC (Hybrid Fiber Coax) network. The Policy Server tracks resources allocated between the CMTS (Cable Modem Termination System) and cable modems. The COPS (Common Open Policy Service) protocol [2], is commonly used between a Policy Server and network elements for setting policies such as QoS and security. The Policy Server uses COPS to manage HFC resource allocations. For VoIP calls either an IMS network is used or a Call Application Server. DOCSIS 1.0 was defined prior to IMS and specified using MGCP as its signaling protocol. In DOCSIS 2.0 the signaling architecture moved to IMS and uses SIP as its signaling protocol.

8.2 Integrated Services

Integrated Services, also known as IntServ, is an approach to offer end-to-end QoS for IP streams. IntServ uses a signaling protocol, most notably RSVP [3], described in the next section, to signal QoS traffic parameters hop by hop across an IP network. The purpose of signaling hop by hop is to establish resources at each network element along the path for an IP stream being established. Reserving and allocating resources at each network element allows the requested traffic and QoS to be honored. If any network element along the stream's path cannot honor the requested traffic and QoS, then the reservation is rejected. Traffic and QoS parameters are captured using flow specs, which consist of a TSpec and an RSpec. TSpec stands for traffic specification, which defines the amount of traffic bursts and how much bandwidth is needed for the stream or flow. The TSpec contains traffic parameters such as Peak Data Rate, a token bucket size, and a token bucket drain rate. Token buckets characterize the flow of traffic that can be sent within a particular time period. The RSpec stands for reserve specification, which defines the desired service in terms of best-effort service or a service requiring minimum delay and jitter guarantees.

IntServ is an IP architecture with the key aspect that resources are reserved at all networking elements along a path for each individual flow or stream. The benefit of IntServ is that detailed QoS can be established on a per-flow basis. One major obstacle for IntServ is its ability to scale in large networks that can have hundreds of thousands of IP flows.

8.2.1 RSVP

RSVP (Resource Reservation Protocol) is the signaling protocol used by the IP IntServ QoS architecture. RSVP is defined in IETF RFC 2205 [3]. Two primary messages are used by RSVP to establish and refresh QoS reservation flows. The two messages are Path and Reservation (Resv) messages. The base RSVP protocol is intended for unicast and multicast flows. A unicast flow is a single-directional flow from sender to receiver. A multicast flow is from one sender with multiple receivers, which is useful in video broadcast applications. The Path message is sent from the sender to the listener(s). The Path message contains the IP address of the previous node so that it can propagate messages like the Resv in the reverse direction of the flow. Also included in the Path message is the sender's TSpec that defines the traffic characteristics being proposed and a sender's template that describes the format of the sender's data. The Resv message is sent from the receiver to the sender along the same path in the reverse direction of the Path message. The receiver initiates sending the Resv message after it has processed the Path message. The Resv message contains the flow spec information such as the RSpec that defines the QoS parameters to be used for the stream.

Figure 8-2 shows the RSVP messaging for establishing, refreshing, and tearing down a reservation flow.

- 1–4. The RSVP sender initiates a RSVP reservation to the receiver by sending a PATH message that contains a TSpec.
- 5–8. The RSVP receiver processes the PATH message and accepts the request by sending a RESV message back to the sender. The RESV message contains the flow spec specifying the QoS parameters for the flow. Once the sender successfully receives

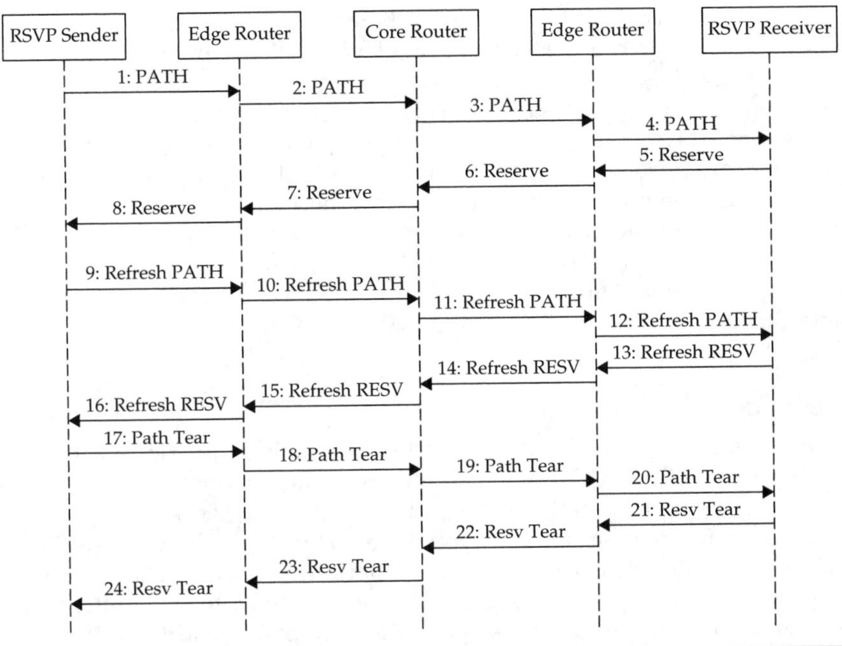

FIGURE 8-2 RSVP messaging

the RESV message, the flow has been completely reserved in all routers. At this point the requested traffic and QoS performance should be available for the IP flow from the sender to the receiver to use.

9–12. After a refresh interval of time has past, usually on the order of minutes, if the QoS guarantee flow is still needed, the sender will refresh the reservation by resending the PATH message. The refresh PATH message is a replica of the original PATH message sent in message 1.

13–16. When the RSVP receiver processes the refresh PATH message, it can decide to continue the reservation by sending an RESV message back to the sender. The RESV message is a replica of the original PATH message sent in message 5.

17–20. At some point, the sender no longer has anything to stream to the receiver. As a result, the sender initiates freeing up the reservation resources by sending a PATH Tear message. Note that if the sender just stops streaming and didn't send the PATH Tear message, the soft state would age out all allocated resources for the flow from sender to receiver.

21–24. Once the receiver receives the PATH Tear message, it frees its allocated resources and sends a RESV Tear in the reverse direction. All routers along the path will free up their resources as they process the PATH Tear message.

RSVP uses a concept of soft state, where the path and reservation state is required to be refreshed on a regular basis. To refresh the state, the Path and Resv messages are exchanged end to end as if the resources haven't been allocated already. If the Path and Resv state is not refreshed within a refresh period, then the allocated resources for that path are freed up in each network element. Although RSVP does have a tear-down procedure, the soft state mechanism ensures that valuable network element resources do not go stale as a result of tear-down error scenarios. The one main disadvantage of a soft state is the scaling of refreshes with large numbers of flows in networking equipment without imposing significant processing load.

8.3 DiffServ

Differentiated Services (DiffServ) is an IP QoS architecture designed to improve on the scalability issues of IntServ. IntServ requires every QoS flow to signal a path end to end through the IP network and maintain a reservation state. For very large networks with hundreds of thousands of flows, scaling becomes a performance issue. DiffServ takes an approach where IP traffic can be aggregated into QoS traffic classifications. IP traffic is classified at the edge of an IP administrative domain where each IP packet is labeled with a six-bit Differentiated Services Code Point (DSCP). The DSCP is inserted in the Type of Service (ToS) field for IPv4 packets. Figure 8-3 shows the ToS field in an IPv4 packet header. Prior to DiffServ and DSCP code points, the original definition of the IP ToS field is shown in Figure 8-4. For IPv6 the DSCP values are carried in the Traffic Class field. Figure 8-5 shows Traffic Class field in an IPv6 packet header. At each router the DSCP value is used to process the packet with a priority that is called the Per-Hop Behavior (PHB).

Using a six-bit encoding allows for 64 different traffic classifications. Several RFCs define encoding values for DSCPs and how they should be used in an IP network. Figure 8-6 shows the general format

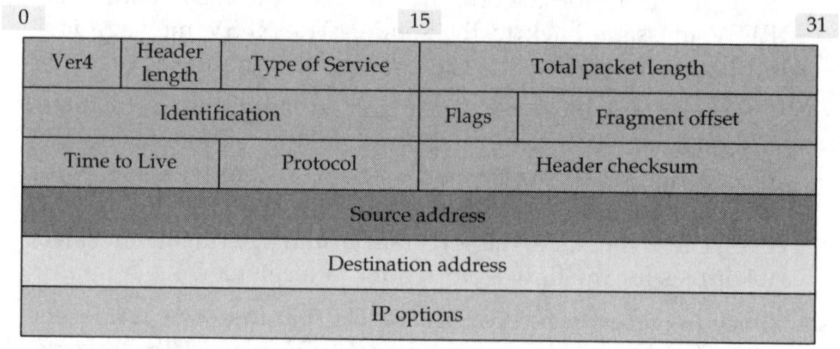

0			15		31
Ver4	Header length	Type of Service	Total packet length		
Identification			Flags	Fragment offset	
Time to Live		Protocol	Header checksum		
Source address					
Destination address					
IP options					

FIGURE 8-3 IP Version 4 packet header

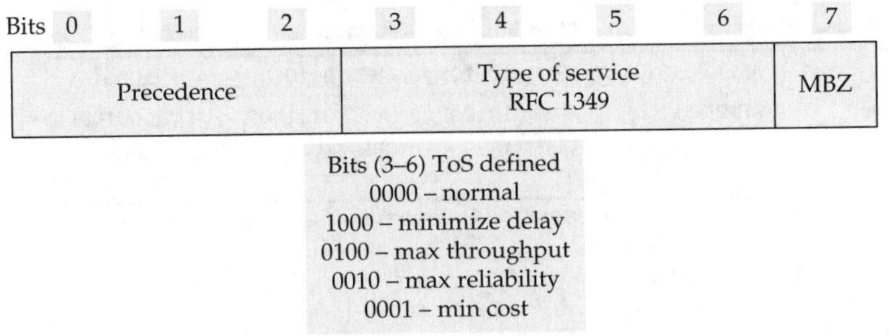

FIGURE 8-4 ToS field prior to DiffServ

for the DSCP field provided by RFC 2474 [4]. A DSCP value of 0 is the default PHB, which is used for any traffic that doesn't match a classification and is treated as best effort. This is the first class of service that will be dropped if any type of congestion occurs in the network. RFC 3246 [5] defines an Expedited Forwarding (EF) PHB, which is used for flows that require low loss, low delay, and low jitter performance. Expedited Forwarding is given the highest priority over all other services. This would mostly be used for real-time interactive applications such as voice or video conferencing.

RFC 2597 [6] defines the Assured Forwarding (AF) behavior. The intention of Assured Forwarding is that packets will be forwarded provided the flow rate is within the rate that was provisioned. When the flow rate exceeds the provision rate the excess packets will be treated as best effort. AF is good for services such as data transfers. The packet order is also intended to be preserved. AF defined four classes with three levels of drop precedence. When congestion occurs,

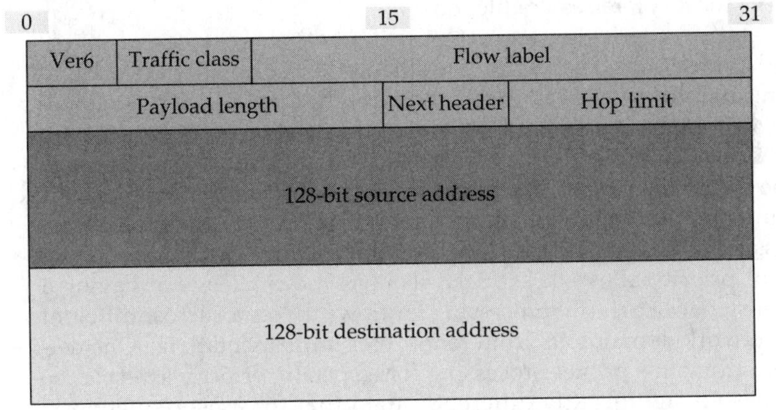

FIGURE 8-5 IP Version 6 packet header

Drop Precedence	Class 1	Class 2	Class 3	Class 4
Low Drop	AF11(001010)	AF21(010010)	AF31(011010)	AF41(100010)
Medium Drop	AF12(001100)	AF22(010100)	AF32(011100)	AF42(100100)
High Drop	AF13(001110)	AF23(010110)	AF33(011110)	AF43(100110)

TABLE 8-1 Assured Forwarding Classes and Precedence

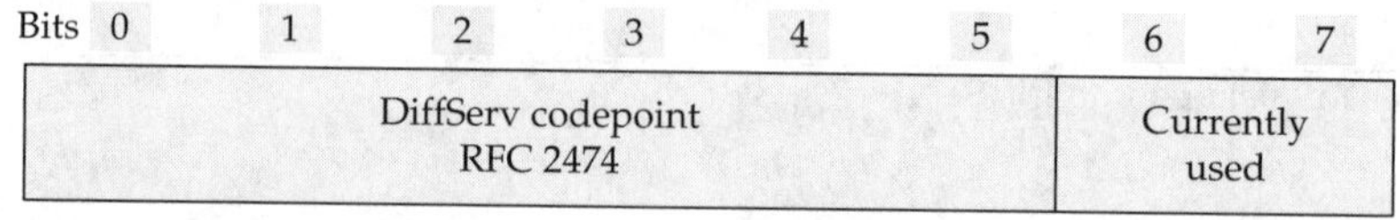

FIGURE 8-6 DiffServ DSCP field

flows that match a higher drop precedence will be dropped before the flows that are in a lower precedence. The four classes will also have differences of priority. With this approach there are essentially 12 different levels of precedence. Table 8-1 shows the four different classes and the three drop precedences within each class.

8.4 IntServ and DiffServ Compared

IntServ provides per-flow QoS establishment by first signaling hop by hop across an IP network. This allows IntServ the ability to provide very finely detailed levels of QoS from one flow to the next. The disadvantage is the ability to scale in very large networks. A significant part of the scaling issue is that RSVP requires the soft state to be refreshed. In order to address this issue, RFC 2961documents mechanisms for RSVP refresh reductions [7].

DiffServ provides for a QoS differentiation among different classes of service. Packets are classified at the edge of a network and don't require a reservation signaled hop by hop. This approach allows DiffServ the ability to overcome the scaling problem with IntServ. As a result, DiffServ is far easier to implement and maintain. One disadvantage of DiffServ is that within a class of service there is a limit to the priority resolution among all of the streams that get classified the same. As seen in Assured Forwarding, there are 12 levels of priority allowed. DiffServ also has the challenge of having a consistency of priority processing for flows that span across different IP network domains and different router implementations. One network domain's packet processing for a specific priority level can be processed significantly differently than that for a peer's network, resulting in inconsistency of QoS performance to the user.

8.5 Layer 2 QoS

Layer 2 switching is generally performed by looking up a tag or label to forward packets to the next switch or endpoint in the network. The layer 2 label is only unique at the data link layer. In comparison, layer 3 routing and switching use addresses such as IPv4 or IPv6 that are network unique. Two types of well-known layer 2 switching that support QoS classifications are Frame Relay and ATM. Frame Relay specifications can be found in the Broadband Forum [8]. ATM specifications can also be found in the Broadband Forum [8]. ATM defines a signaling interface between the user equipment and network called the User Network Interface (UNI). ATM also defines a signaling and routing interface used between peer networking equipment known as the Network-to-Network Interface (NNI). ATM uses a signaling protocol that is derived from the ISDN signaling protocol called Q.2931 [9]. For routing, ATM uses a link state protocol called PNNI.

Both ATM and Frame Relay use virtual circuits by programming packet label mappings at each switch that the virtual circuit traverses. ATM uses fixed-size packets called *cells*. ATM cells are always 53 bytes in length. When a packet on a network node is received on an interface, a label in the packet is used to look up what outgoing port to forward the packet to. When the packet is forwarded, the outgoing packet has its label changed based on the programmed label mapping. In Frame Relay the packet label is called Data Link Channel Identifier (DLCI). ATM's label consists of two parts called the Virtual Path Identifier (VPI) and the Virtual Circuit Identifier (VCI). Virtual circuits can be established either as permanent virtual circuits (PVCs) or switched virtual circuits (SVCs). SVCs are established using a signaling protocol and are torn down when not needed. PVCs are provisioned using a network management system and persist far longer than SVCs. Today most Frame Relay and ATM networks use PVCs.

Frame Relay virtual circuits are created with traffic contracts to allow the proper resources to be allocated and set thresholds to prevent congestion as best as possible. One key traffic parameter is the Committed Information Rate (CIR), which is the average bandwidth the network will allow without dropping packets. Another traffic parameter is the Excess Information Rate (EIR), which defines a burst of traffic the network will attempt to carry provided the bursts are infrequent. Burst traffic packets are marked so that they can be dropped in the network if congestion occurs. Frame Relay also has congestion notification capabilities where the network can send indications to the edge nodes to lower their rate if necessary in order to maintain the expected QoS.

ATM is cell-based switching technology that was designed with QoS in mind. A cell can be thought of as a fixed-size packet. In ATM all cells are 53 bytes in length. Media such as voice, video, or data are supported using the various ATM Adaptation Layers (AALs). AAL1 is used for Constant Bit Rate (CBR) services such as voice and video. AAL5 is used for carrying packets for applications like data. A device known as a SAR (Segmentation and Reassembly) is used at ATM edge nodes to convert media into cell streams and forward it to core switches. Core ATM switches primarily forward cells based on their programmed label mappings. ATM supports programming of two delay QoS metrics at each switching element in the network, which are

- **Delay** At each switch the maximum amount of delay that a switch adds to forwarding cells for each virtual circuit is programmed. This allows for the end-to-end delay of a virtual circuit to be constrained to a desired value, as is very important for real-time applications. ATM uses a prioritized queuing mechanism in its switch fabrics to honor the various delay requirements of all VCs programmed.
- **Delay variation** At each switch the maximum delay variation a VC should experience is programmable. Delay variation is essentially the cell jitter measure for a VC's cell stream. This allows the end-to-end delay variation of a virtual circuit to be constrained to a desired value, as is also very important for real-time applications. ATM uses traffic shapers as described in the section "Traffic Controls" of this chapter to achieve the desired delay variation.

ATM also uses traffic policing as described in the section "Traffic Controls" of this chapter in order to prevent network congestion occurrences.

8.6 CAC

Non-real-time applications, such as data traffic, are usually sent with a best-effort QoS. If a packet gets delayed or dropped, due to network congestion, the data application most of the time can recover without any impact to the user. Real-time applications are much more vulnerable to network congestion. Significant increases in end-to-end delay or packet loss can be very noticeable and disruptive to the user of real-time applications such as voice or video. These performance degradations can occur if too much real-time traffic is admitted into network, causing network congestion. In order to avoid this type of overbooking, a mechanism known as Call Admission Control (CAC) is imposed. CAC is a comparison calculation between the known available network resources and resources needed to honor a newly requested stream. If there are enough available resources to add the newly requested stream, then the CAC algorithm allows the new stream to be established. If there aren't enough spare resources for the new stream, then the CAC algorithm dictates that the new request

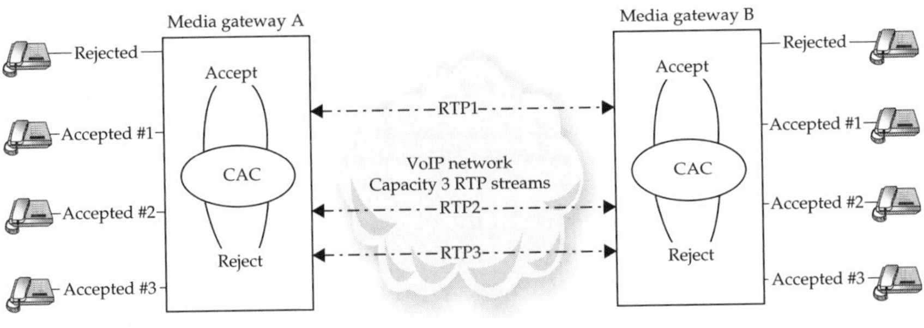

FIGURE 8-7 CAC example

should be rejected for the sake of not impeding the current real-time streams in the network. Figure 8-7 is a simplified example of CAC being used by two media gateways. In this example the maximum capacity of the VoIP network is three voice RTP streams. If any additional RTO streams are added, then the performance of all voice calls will be degraded. The CAC algorithm in the media gateways simply tracks the maximum network capacity and as shown rejects a fourth voice conversation being requested.

There are several categories of CAC. In the example of Figure 8-7 the media gateways did not allow more than three voice calls based on provisioned configuration information. This form of CAC is known as local CAC, since all information used to form a CAC decision is based on local information. If in the CAC example RTCP (RTP Control Protocol) packets were sent regularly for each of the RTP streams to obtain current performance statistics, then the CAC algorithm could use that additional information as a variable in the CAC decision process. Using measured information such as RTCP statistics data is called *measurement-based* CAC. Another type of CAC that can be used is known as *resource-based* CAC. In resource-based CAC the resources of the network are inventoried, as opposed to the local CAC, which inventories the resources of a single local node. Some of the resources that are tracked are bandwidth and switching capacity. Also included are hardware limits of networking equipment such as trunk time slots, network processor capacity, DSP limits, CPU processing abilities, and memory available. Resource-based CAC can be performed from a centralized server such as a Policy Server. Another approach is to use network-based CAC at the border of a network such as a P-CSCF or Session Border network equipment used in IMS.

8.7 Traffic Controls

In order to offer QoS guarantees, traffic controls are required at different points in the network. The main traffic control categories are traffic priorities, policing, and traffic shaping. Using traffic priorities allows higher-priority traffic to be forwarded ahead of lower-priority traffic. This allows for a different class of service where real-time applications can be assigned a high priority, resulting in needed lower latencies. Lower-priority traffic, which can be assigned to applications such as data, would have higher delays and jitter without any implications to the end user.

Traffic policing monitors the traffic of individual streams, verifying that the rate doesn't exceed the streams' traffic contract allocation. If a stream's traffic approaches or exceeds its traffic contract, various controlled actions can be taken. Some of those actions are as follows:

- **Packet discarded** When a burst of traffic on a stream significantly exceeds the stream's traffic contract, packets over the contract rate can be discarded. Dropping packets is the most extreme action taken, since it could be noticeable by the application if the percentage of packets discarded is significant.
- **Mark packets** If there is enough capacity on the outgoing link for a node, an alternative to discarding packets is to mark the outgoing packets as eligible for dropping. When a downstream node receives packets that are marked, it will discard those packets first in the event it determines it needs to drop packets. Packets are marked if the flow rate is slightly above the traffic contract thresholds. With this approach, the network can be more forgiving by allowing some bursts of traffic above the average traffic flow without taking the extreme action of discarding packets.
- **Flow control indications** When network congestion occurs at a given node, indications can be sent upstream and downstream to inform the stream source to reduce its rate. This allows for a means of flow control, thereby avoiding long periods of congestion.

Traffic shaping is a widely used traffic control mechanism to make sure the QoS and traffic contracts are honored. Traffic shaping controls the maximum traffic sent from a source over a specific time interval. By controlling the packet rate sent from a networking device, the bandwidth and the bursts of traffic can be maintained at deterministic levels. There are two well-known traffic shaping algorithms, called leaky bucket and token bucket algorithms.

Networking equipment that processes streams of varying rates and burst intervals, without traffic shaping, can send/forward packets that have large variances in rates. Figure 8-8 shows a packet stream feeding into the top of a leaky bucket that contains a couple of packet bursts. The bucket, which is essentially a First In First Out (FIFO) queue, absorbs the burst of packets into the bucket. The output of the bucket is a constant drip of packets that is smooth with the bursts

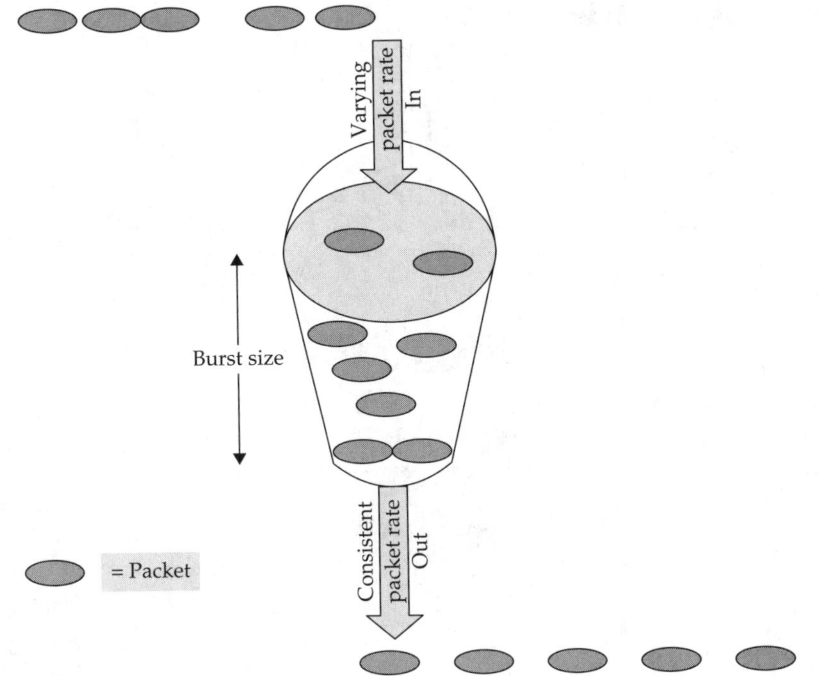

FIGURE 8-8 Leaky bucket algorithm

removed. A leaky bucket will empty the FIFO, at a configurable interval that sets the maximum packet rate. The larger the bucket or FIFO size, the larger the burst the leaky bucket can absorb. However, as the FIFO size is increased, so is the maximum delay introduced by the leaky bucket. So there is a trade-off to be made between the potential delay added and the maximum burst size a leaky bucket can absorb. Also note that if the average arrival rate exceeds the configured FIFO drip rate, the bucket will overflow causing packets to be discarded. The end result of using a leaky bucket is that the traffic leaving the bucket will be smooth without bursts or jitter, as well as have its bandwidth limited to a deterministic value.

The token bucket algorithm takes a different approach from the leaky bucket algorithm. The token bucket algorithm is a generic rate comparison algorithm that can be used for traffic policing, traffic shaping, or some other traffic control mechanism that can be based on a rate comparison. Instead of completely removing all bursts of traffic, the token bucket will allow for a short duration of bursts, provided the long-term average traffic rate doesn't exceed the allocated traffic contract. Figure 8-9 is a diagram of a token bucket. Tokens are deposited into the bucket at a fixed rate. Once the token bucket is filled, no more tokens can be added. In the token bucket algorithm, usually based on packet size, a certain number of tokens are needed to be present in the

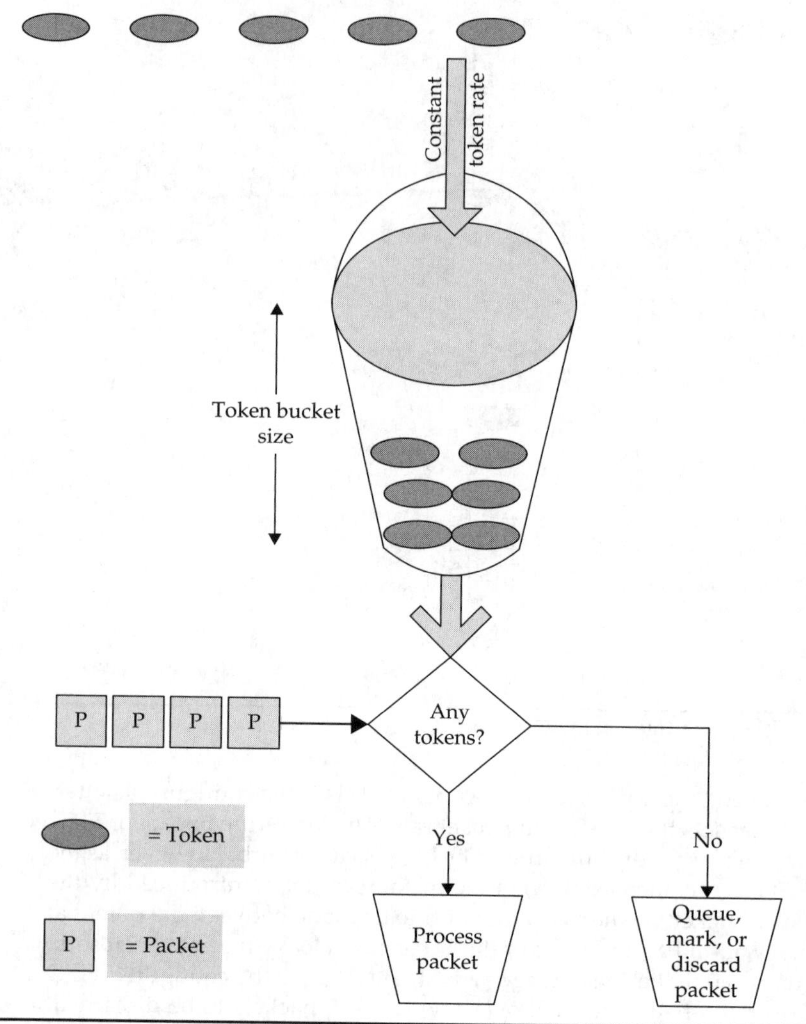

FIGURE 8-9 Token bucket algorithm

bucket in order to proceed with processing each packet. For an example, one token could be required for every 100 bytes of a packet. When a packet arrives to be processed, a decision is made as to whether there are enough tokens to process that packet. As shown in Figure 8-9, the result of the decision yields the following outcomes:

- If there are enough tokens in the bucket, the tokens are removed from the bucket. The packet is then processed. If the token bucket is being used for shaping, the packet is transmitted. If the token bucket is being used for policing, then the packet is considered accepted and continues to be processed.

- If there aren't enough tokens in the bucket and the token bucket is being used for shaping, then the packet will be queued until enough tokens are placed into the bucket to proceed with processing the packet. Note that if the traffic shaper queue is already full, then the packet will be discarded. If there aren't enough tokens in the bucket and the token bucket is being used for policing, the packet is either discarded or marked as eligible to be discarded.

8.8 GPRS QoS

GPRS (General Packet Radio Service) is a packet technology introduced in 2G wireless networks to significantly increase the data bandwidth available to wireless endpoints. Before GPRS the data traffic was very limited for wireless devices and occupied the same circuit-switched channels that voice is carried on by using an analog modem. GPRS packets are sent on a channel dedicated to data purposes. On the high end, GPRS allows for speeds of 50 Kbps. Many of today's mobile applications can be supported using GPRS, such as Multimedia Messaging Service (MMS) for sending photographs, Internet Access, Instant Messaging, and Push to Talk walkie-talkie service.

The original GPRS, like other packet technologies, uses a best-effort delivery of packets. With IP networks offering QoS using IntServ or DiffServ, GPRS networks are able to leverage these capabilities to offer QoS classes as well. Figure 8-10 shows a basic GPRS IP Access diagram. Mobile terminals send IP packets using a dedicated packet tunnel to a GGSN (Gateway GPRS Support Node). A GGSN is responsible for terminating multiple tunnels with mobile terminals and routing IP packets to one or more packet networks. Two segments

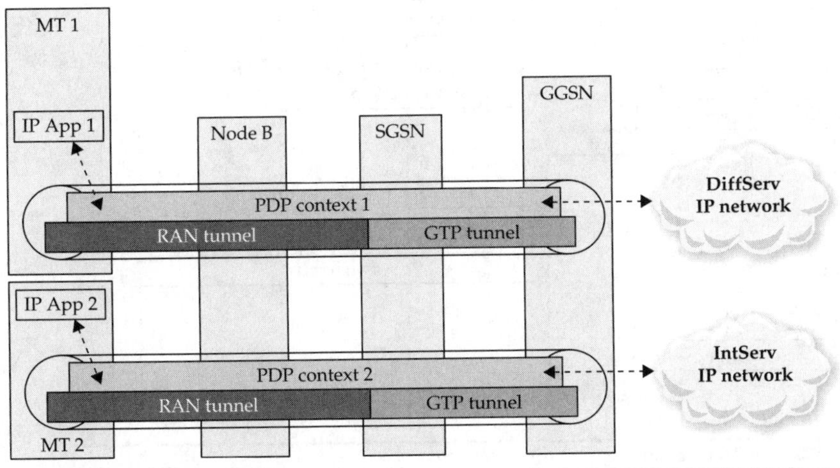

FIGURE 8-10 GPRS IP access

exist that make up the GPRS packet tunnel. The first segment is between the mobile terminal and the SGSN (Serving GPRS Support Node). The SGSN is responsible for mobility management, access control, and security. The second segment of the tunnel is between the SGSN and the GGSN that uses the GTP (GPRS Tunneling Protocol). Mobile terminal applications need to establish a PDP (Packet Data Protocol) context with the GGSN before it can send and receive IP packets. PDP context attributes include the following:

- PDP type and address
- QoS parameters profile
- Authentication information

Figure 8-11 shows the sequence diagram to establish two separate PDP contexts from one mobile terminal. The first PDP context will be used for sending packets over a DiffServ IP network. The other context will be used for sending packets of a IntServ IP network. In this example two different IP networks are used, each supporting a different range of QoS performance characteristics.

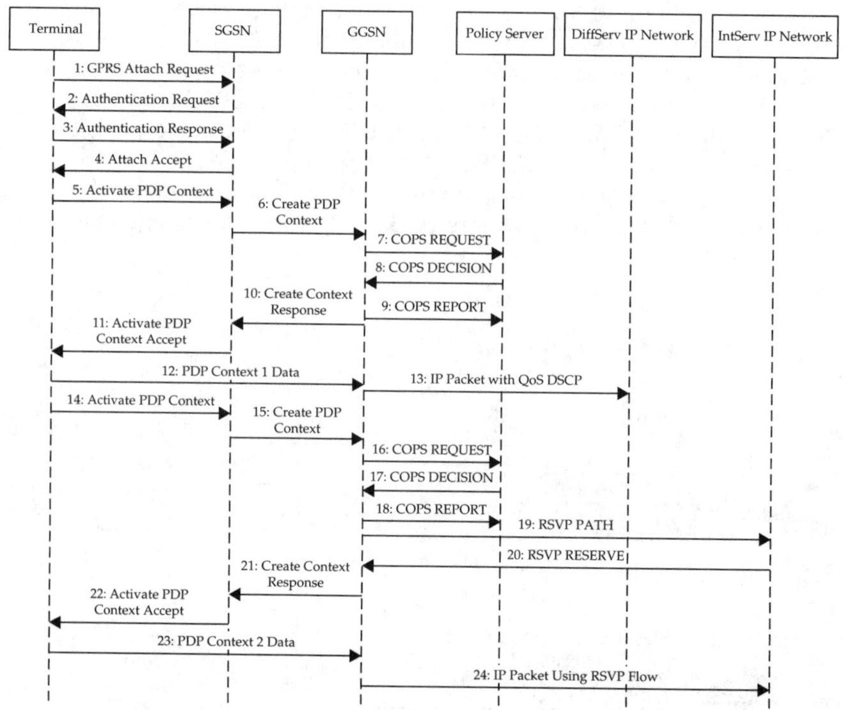

FIGURE 8-11 GPRS QoS flow

The following is a description of the message flow from Figure 8-11.

1. The mobile terminal first attempts to attach to the SGSN.
2. The SGSN challenges the mobile terminal by requesting authentication.
3. The mobile terminal sends its authorization information back to the SGSN.
4. The SGSN accepts the attachment and is now ready for a PDP context to be opened.
5. The mobile terminal sends a PDP activate context request specifying the QoS parameters needed for the context.
6. The SGSB in turn sends a PDP create context to the GGSN with the QoS parameters needed.
7. The GGSN sends a COPS (Common Open Policy Service) Request to the Policy Server that checks if there are enough resources to honor the request.
8. After the Policy Server performs CAC, it approves the request by sending a COPS decision message back to the GGSN.
9. The GGSN informs the Policy Server of the amount of resources consumed by sending a COPS report message to the Policy Server.
10. The GGSN accepts the context by sending a create context response back to the SGSN.
11. The SGSN in turn sends an Activate PDP context Accept message to the mobile terminal informing it that the context has been created and is ready to use.
12. The mobile terminal IP application sends a packet over the PDP context.
13. The GGSN forwards the packet over the DiffServ IP Network using a DiffServ code point that satisfies the QoS profile for the GRPS context.

Messages 14–24 establish another GPRS context that uses an integrated services IP network. A RSVP path is established by the GGSN to reserve the QoS resources in the IP network to comply with the requested QPRS context.

References

[1] DOCSIS (Data over Cable Service Interface Specification) 1.0 Interface Specification.
[2] IETF RFC 2748, "The COPS (Common Open Policy Service) Protocol", January 2000.

[3] IETF RFC 2205, "Resource Reservation Protocol (RSVP) Version 1 Functional Specification", September 1997.
[4] IETF RFC 2474, "Definition of the Differential Services Field (DS Field) in the IPV4 and IPV6 Headers", December 1998.
[5] IETF RFC 3246, "An Expedited Forwarding PHB (Per Hop Behavior)", March 2002.
[6] IETF RFC 2597, "Assured Forwarding PHB Group", June 1999.
[7] IETF RFC 2961, "RSVP Refresh Overhead Reduction Extensions", April 2001.
[8] Broadband Forum, www.broadband-forum.com.
[9] ITU-T Q.2931, "Digital Subscriber Signaling System No. 2 – User Network Interface (UNI) layer 3 specification for basic call/connection control", February 1995.

CHAPTER 9
IMS Network Architecture

IMS (IP Multimedia Subsystem) is a network architecture originally specified by the 3GPP (3rd Generation Partnership Project) wireless standards body designed for supporting multimedia services. Since then IMS has been expanded to include the majority of mobile wireless technologies as well as fixed broadband access technologies. The major benefit of IMS is that it offers a common approach to establishing multimedia connections across mobile and fixed broadband networks. The intention is to converge on a protocol suite allowing interoperability among networks independent of the access to the end user, including DSL, cable, and various wireless technologies.

This chapter will describe the IMS architectural components and their interfaces. Detailed call flows will be provided that show how multimedia services are established in an IMS network.

9.1 IMS History

The first version of IMS was introduced by 3GPP Release 5 for wireless networks. Release 5 IMS offered a common approach to implementing multimedia services through the use of wireless packet access technologies based on GPRS (General Packet Radio Service) for GSM and UTRAN (Universal Mobile Technology System Terrestrial Radio Access Network). The UTRAN air interface uses W-CDMA (Wideband Code Division Multiple Access) rather than its predecessors TDMA (Time Division Multiple Access) and FDMA (Frequency Division Multiple Access).

In Release 6, 3GPP added IMS support for wireless LAN access. Then in 3GPP release 7 support for fixed-line access was added. By supporting wireless mobile, wireless LAN, and fixed line access, IMS provides a path to FMC (Fixed Mobile Convergence). With FMC, a single phone device can be used for placing calls using wireless LAN or wireless mobile technologies. The early implementations of FMC would require dual-mode handsets. A dual-mode handset uses transceivers with two types of wireless technologies such as CDMA and WIFI.

With the advent of femtocells, FMC can be achieved using single-mode handsets, which drastically reduce power consumption as well as provide other significant benefits.

9.2 IMS Components

This section will describe the network components of the IMS architecture. IMS defines components to have specific functions in the network. Commercial network equipment can implement one or multiple IMS functions. There are pros and cons with having multiple IMS functions in a network device. One obvious advantage is that there is less network equipment to manage and maintain. A disadvantage is the reduced flexibility to swap one of the IMS functions using a different supplier's network equipment.

Figure 9-1 shows the various network components that make up an IMS network. This figure is split into two parts, the access network and the IMS core.

9.2.1 Access Network

A customer can connect to an IMS network using various access technologies. The access can be a fixed broadband technology that offers high-speed IP service. The following are some examples of fixed broadband access technologies:

- Hybrid Fiber Coax (HFC) with a cable modem is a typical access technology used by cable TV service providers.

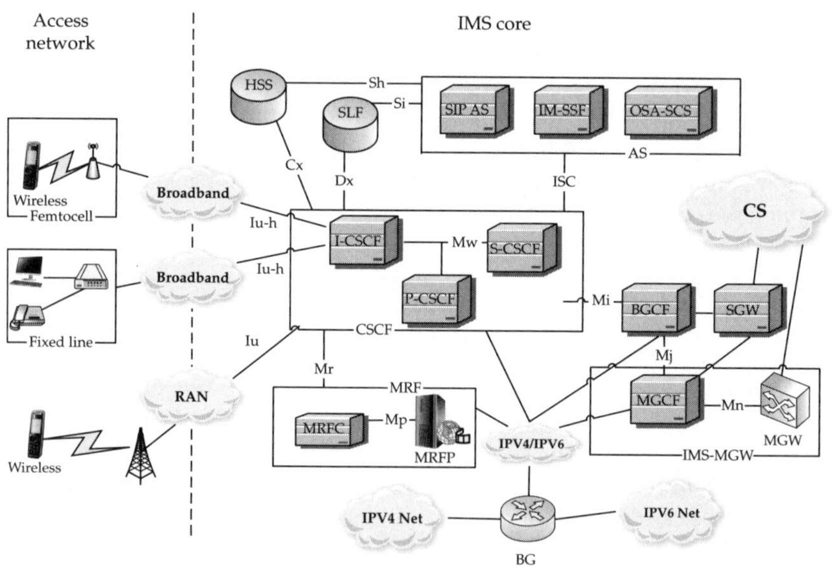

Figure 9-1 IMS network components

- PON (Passive Optical Network)/GPON (Gigabit Passive Optical Network) provides access to the network through a fiber-optic connection that offers the customer many services, including broadband IP.
- DSL (Digital Subscriber Loop) provides broadband IP service using existing telephone copper local loop wiring. Digital Signal Processing is used to allow for full-duplex digital transmission on a single twisted-pair cable. DSL transmission rates are slower then HFC and PON technologies; however, they take advantage of reusing existing wiring between the telephone company and the customer's home.
- WiMax (World Wide Inter-operability for Microwave Access) provides broadband IP access using microwave digital transmission. A radio interface is used for the last mile to the end customer. This service provides a relatively high-speed IP access for customers.

At the customer's home, the device that terminates the IMS SIP protocol could either be the access provider's gear such as a cable modem, a WiMax Access Device, or a DSL residential gateway. If the IMS service is not provided by the access service provider, then an additional device is needed, such as an IAD (integrated access device) that implements SIP on the IP access side and supports media interfaces including a phone line or possibly a video interface. Users can get access to the IMS network through various wireless mobile access technologies. Some examples of mobile wireless access technologies are: GSM, W-CDMA, LTE, and WiMax.

9.2.2 IMS Core

The IMS core network contains all of the network equipment to offer services to the end customer. The service offered can be voice, video, or other applications such as text messaging. The following sections describe the components that make up an IMS core network.

CSCF

The CSCF (Call Session Control Function), as the name suggests, uses SIP to provide a session and call control function. CSCF consists of three types of SIP servers, each needed for different aspects of the SIP signaling within the IMS core. The three types of CSCF servers are as follows:

- P-CSCF (Proxy CSCF)
- I-CSCF (Interrogating CSCF)
- S-CSCF (Serving CSCF)

Proxy-CSCF The Proxy-CSCF is the first SIP server that an IMS endpoint device initiates communications with. An IMS terminal, as shown in Figure 9-1, can be a mobile wireless terminal or a fixed-line device that uses broadband access. In the case of a fixed-line broadband IMS terminal, the P-CSCF is discovered using DHCP or through manual provisioning of the P-CSCF address in the IMS terminal or IAD. A mobile wireless IMS terminal would discover its P-CSCF though DHCP or through the PDP context from the GPRS. Here are some of the aspects of the P-CSCF:

- A P-CSCF is assigned to an IMS terminal during registration. The P-CSCF can be located in the home network or a visited network for wireless devices. When the P-CSCF receives a registration from the IMS terminal, it forwards the registration to the customer's home network's I-CSCF based on the Public Address of Record (AOR) received in the registration message.
- Whenever an IMS device changes its location, it registers again with the nearest P-CSCF.
- The P-CSCF monitors all SIP endpoint messaging and performs traffic policing. This allows the P-CSCF to protect the core network from any misbehaving endpoints that generate too much network traffic. This also includes denial of service attack protection whereby a malicious device intentionally generates a large amount of traffic to disrupt service.
- As part of the registration process an IMS endpoint identifies itself using its public address. In order to ensure an IMS endpoint is the device it claims to be, the P-CSCF authenticates the IMS endpoint on a regular interval as part of the SIP signaling exchange.
- The P-CSCF protects the core IMS network from an unsecured IP network. This allows other IMS network components to communicate over a trusted interface. P-CSCF can use IPsec to communicate with IMS IADs to increase the level of security over an untrusted IP network.
- For mobile phones that have limited bandwidth, SigCom (Signaling Compression) is used between the IMS endpoint and the P-CSCF. For a text-based protocol like SIP, compression can significantly reduce the bandwidth needed. Signaling Compression is defined in RFC 3320 [1].
- The P-CSCF can also participate in the QoS reservation of the access network.
- The P-CSCF will manage registration refreshes from IMS endpoints. This is a critical function in networks where high refresh rates occur as a result of NAT (Network Address Translation) pinhole refreshing. The P-CSCF will absorb the brunt of high registration refreshes, protecting the rest of the core IMS network from this performance burden.

I-CSCF (Interrogating CSCF) The I-CSCF operates strictly within a home network and is responsible for dispatching received SIP requests to the appropriate S-CSCF based on data queried from the HSS. For requests destined for outside of the home network, the I-CSCF forwards the request toward the destination using the IBCF (Interconnect Border Control Function). Here are some of the responsibilities of the I-CSCF:

- During SIP Registration, the I-CSCF forwards the registration to the appropriate S-CSCF based on the IMS device's address of record (AOR) and provisioning information obtained from the HSS. This allows multiple S-CSCF to service an administrative domain.
- The I-CSCF forwards SIP requests received from other networks to the appropriate S-CSCF that was assigned during registration.
- The I-CSCF can forward or reject requests destined to users outside of an IMS network.
- Similar to the other CSCF network elements, the I-CSCF generates accounting records that track the network usage.
- The I-CSCF IP address can be published in an administrative network's DNS server, allowing remote servers to forward SIP messages destined for the I-CSCF's home network.

S-CSCF (Serving CSCF) The S-CSCF, similar to the I-CSCF, also is used strictly within a home network. It is responsible for maintaining SIP registration and call state with IMS terminals that belong to its home network. Based on provisioning information stored in the HSS, the S-CSCF will distribute calls to application servers as needed for additional media and call features. Here are some of the responsibilities of the S-CSCF:

- The S-CSCF is the primary SIP server used to perform session and call control. The basic call and registration state is maintained by the S-CSCF. Subscribers provisioning data is obtained from the HSS using the diameter protocol. SIP session timers and call progress states are tracked and maintained by the S-CSCF.
- Depending on the subscriber data and SIP session data received for a call, the S-CSCF can redirect or transfer a SIP dialog to an application server for added feature or media processing. An example of this could be a voice or video conferencing Application Server. Other examples of Application Server usage are provided in the section "Application Server."
- SIP registrations are maintained by the S-CSCF, allowing it to track the Address location of the IMS terminal based on its

Public Address of Record. As registrations and other SIP messages occur, the S-CSCF validates the authorization information against the user profile retrieved from the HSS.

- Routes are obtained from the HSS after performing ENUM (Electronic Number) number translations.
- Several S-CSCF servers can reside in a home network. As the number of IMS customers for a home network increases, more S-CSCFs can be added. An S-CSCF is assigned an IMS terminal during registration as a result of the I-CSCF querying its home HSS. How S-CSCFs get assigned as registrations occur is usually a configurable option with a load-balancing algorithm to prevent any one S-CSCF from becoming overloaded. The I-CSCF keeps track of available S-CSCFs in a home network, allowing the network to continue to process calls in the event of an S-CSCF failure, provided there is another S-CSCF available. This approach allows for an increase in service availability by providing a higher fault tolerance to equipment and network failures.
- The Diameter protocol is used to exchange subscriber profile data between the S-CSCF and the HSS. Like the other CSCF functions, the S-CSCF also generates accounting records as the call progresses.
- Depending on how a call progresses, the S-CSCF can further route a call to another S-CSCF in its home network, redirect the call to the PSTN, or redirect it to a different destination home network using the IBCF.

HSS

The HSS (Home Subscriber Server) stores the subscriber database. All authentication and authorization information, including security keys, needed to validate an IMS terminal during registration is persisted in the HSS. The HSS is similar to the HLR (Home Location Server) in a GSM network in that they both store location information. The HSS maintains data on which S-CSCF is currently serving a registered subscriber.

Besides storing subscriber location information, the HSS could store user service data. It is also possible to have user service data stored in an application server database as well. User service data for a customer can define how incoming calls should be handled. For example, when a subscriber is involved in a voice conversation there can be several options for handling an incoming call such as call waiting or call forwarding to voice mail. Service providers offer many feature options that require provisioning data that can be stored in the HSS's or application server's database for easy additions, deletions, modifications, and retrievals.

Here are some other examples of well-known features that require user data that can be used by either an S-CSCF and or an application server:

- Call Forwarding Data would include the telephone number of where to forward the call. Other options could specify under which scenarios calls should be forwarded, such as for busy calls, no-answer calls, or simply all calls.
- Call Acceptance and Call Rejection Call Lists contain phone numbers that the subscriber can activate to be used to either explicitly accept or reject incoming calls originating from those phone numbers.
- Call Waiting, Three-Way Calling, and Do Not Disturb data can be as simple as either being activated or not.

SLF

As service provider networks grow, the amount of user data stored in the HSS grows. Large networks require multiple HSSs to store all of their data. The SLF (Subscriber Locator Function) is used to look up which HSS stores a given user's data based on its public address. The SLF and the HSS support a diameter interface. As the I-CSCF or S-CSCF process SIP registrations, they will query the SLF first to obtain which HSS the subscriber data is stored on. Once a subscriber has successfully registered, the S-CSCF will save which HSS is associated with a subscriber so that subsequent SIP transactions will not require a lookup by the SLF to find a HSS. IETF RFC 4457 defines a P-User-Database SIP header that allows the I-CSCF to pass the address of the HSS for the registering subscriber to the S-CSCF saving a SLF query. The following summarizes the number of SLF queries needed during SIP registration.

- A single HSS network doesn't require any SLF function since I-CSCF and S-CSCF are provisioned with a single HSS address.
- Multiple HSS network requires a SLF query by both the I-CSCF and the S-CSCF.
- Multiple HSS network supporting the P-User-Database SIP header requires a single SLF query by the I-CSCF. The S-CSCF gets the HSS address from the I-CSCF passed along in the P-User-Database header of the Registration message.

Application Server

An application server in an IMS network extends and enhances features to IMS users. An API (application programming interface) implemented by application servers is used by service providers to integrate and build customized features and applications. These APIs can be proprietary or based on a standard. Advanced services can be built using these APIs; for example, a parent could possibly find the location of her child's mobile phone via a web portal. Another application would be to have a web-based GUI to place outgoing and manage incoming calls interactively. Here are two examples of well-known 3PCC (3rd Party Call Control) standards:

- Parlay X [2] is a web-based service API to build third-party applications for fixed and mobile phone networks. Parlay X Web Services are specified by ETSI, 3GPP, and the Parlay Group. The OSA-SCS will use the Parlay API to offer an application interface to an IMS network.
- JAIN (Java APIs for Integrated Networks) [3] supports numerous APIs, including protocol support for SIP and TCAP. It also supports building third-party call control features using Java.

Application servers are used to build advanced features over and above a basic call within an IMS network. The application server functionality, if not built into the S-CSCF, will interface to the S-CSCF using SIP. Here are a few example features application servers can implement:

- VCC (Voice Call Continuity) [4], allows calls to be moved between conventional mobile circuit-switched networks and an IMS packet-switched network. The IMS network can be a wireless access network or a fixed broadband access network. This allows wireless devices to gracefully hand off between fixed-access networks and mobile networks as a user enters or leaves a building, leveraging the most desired connection for the current location.
- Web Portal Call Control is an application where the user can use a web interface to initiate a call using phone numbers from a host database. The Web portal will also allow for interactive call management where pop-ups occur as calls arrive for the user, giving that user a choice as to how to handle the call. Based on the caller ID in the pop-up, the user can decide to redirect the call to voice mail, send the call to one of her SIP devices (home or mobile), or immediately reject the call. The user can access all these capabilities from her favorite web browser, gaining access to managing the stored phone numbers and customizing automatic call handling rules.
- Called Party Locator is a feature that allows an application server to store a list of telephone numbers where a subscriber can be reached. For example, a customer can configure the feature to have his home number, his work number, and his mobile phone number. One option would be to have all phones ring at once when someone calls the subscriber or search through the list one number at a time with a configured number of rings for each number attempted.

SIP-AS A SIP-AS (SIP Application Server) operates in the service provider's home network and has access to the HSS subscriber's database using the Diameter protocol. A SIP-AS interfaces to CSCFs using SIP for processing calls. The SIP-AS decides how a call should be processed such as a basic call, call forwarding, or possibility redirecting the call to the IMS network. Like other application servers, the SIP-AS provides enhanced services beyond the basic call between two IMS users. A SIP-AS can use one of three styles of SIP protocol usage:

- A UA (User Agent) implementation is used by the application to either originate a call or accept incoming calls. A simple application would be for users to dial in to a voice menu system that could be used for any kind of transaction processing such as trading stocks or purchasing tickets.
- A B2BUA (Back to Back User Agent) is a network device that manages two legs of a call. On one leg an incoming call is received via a SIP INVITE, and the B2BUA will maintain it as one dialog. The B2BUA processes the INVITE request and, based on the provisioning data retrieved from the HSS, determines how the call should be handled. Some options for the B2BUA are to forward, transfer, or release the call. The B2BUA can implement advanced features like the called party locator feature, where each location attempted would be another leg of the call.
- A SIP Proxy receives a SIP INVITE and in general forwards the SIP endpoints request toward the called party destination. One main distinction between a SIP Proxy and a B2BUA is that a single SIP dialog is used for a SIP Proxy, whereas more than one dialog are used for a B2BUA. As a result, a SIP Proxy is more limited in its call features than a B2BUA.

IM SSF

An IP Multimedia Service Switching Function (IM-SSF) is used to interface with CAMEL (Customized Applications for Mobile Networks Enhanced Logic) networks. CAMEL is a collection of standards defined by ETSI designed to offer IN (Intelligent Network) features for wireless devices including GSM and UMTS. Intelligent Networks offer enhanced telephone services to customers based on Signaling System 7 (SS7) infrastructure that has evolved through the efforts of the traditional telephone service providers. Some of the enhanced features include: 800 calls, number portability, phone voting, call filtering, and calling cards.

The IM-SSF interfaces to CAMEL networks using the Camel Application Part (CAP) protocol [5] on its SS7 interface and manages a SIP session on the IMS side. The function of an IM-SSF application server is to map SIP sessions to CAMEL SS7 sessions seamlessly.

OSA-SCS

The Open Service Access–Service Capability Server (OSA-SCS) Application Server provides a programming interface based on the OSA framework defined by 3GPP and ETSI. This is a standard API based on the defined Parlay API. An OSA-SCS will use the Parlay API to offer an application interface to an IMS network. Parlay offers a way to implement web-based third-party call control applications.

MGCF

A media gateway control function (MGCF), as the name implies, controls media gateways (MGWs). On one side, the MGCF interfaces to other IMS devices using SIP. On the other interface, the MGCF interfaces to MGWs using a standard VoIP signaling protocol. The VoIP signaling protocol can be H.248, MGCP, or SIP. H.248 is the preferred protocol because of its flexibility and extensibility of control of various circuit switch technologies.

The MGCF also supports signaling to an SS7 ISUP network [6]. This allows an IMS network the ability to place calls to a PSTN (public switched telephone network) using a circuit-switched connection and SS7 signaling. The MGCF uses SCTP at the lower-layer protocols to carry SS7 signaling over IP.

In some implementations the MGW, MGCF, and SGW functions are integrated into one device. This allows for fewer interfaces to maintain in the IMS network.

BGCF

The breakout gateway control function (BGCF) is a SIP server that is responsible to route IMS calls that are destined for a circuit switch network. The BGCF uses the called-party telephone number to perform a lookup that results in choosing an outgoing circuit-switched network to continue establishing the call. The destination circuit-switched network can be either a PSTN (public switched telephone network) or a public land mobile network (PLMN).

SGW

The signaling gateway (SGW) provides a lower-layer protocol conversion for an IMS IP network to an SS7 signaling network. On the IMS interface signaling messages are received from the MGCF that are ISUP and use SCTP and IP as the lower-layer protocols. On the SS7 network side the protocol stack uses ISUP with conventional SS7 lower-layer protocols such as MTP2 and MTP3 [7]. The SGW allows the MGCF to do its job of mapping SIP sessions to SS7 sessions.

MRF

The media resource function is responsible for providing value-added media streaming support. A very common need in a voice network is to offer announcements to provide explanations of problems or instructions to the caller. For example, if a user dials an invalid number, an announcement can be played that instructs the caller to check the number and informs her that she dialed an incorrect number. There are many such scenarios where announcements provide useful information to the caller.

Another commonly used feature offered by a MRF is the ability to bridge together voice streams from multiple endpoints. This capability is used to offer various types of conferencing. In this case the RTP streams are usually converted back to analog voice, then mixed together, and then encoded back into the multiple RTP streams.

Interactive voice response (IVR) capabilities are also implemented in the MRF. IVR allows a call to be placed in an audio dialog where the caller can be prompted for input. One example here would be a caller dialing the phone number of his bank and, after entering his account number and PIN, choosing from a menu to get the checking account balance or savings account balance or possibly transfer funds. IVR dialogs are supported by MRFs using a standard known as VXML (Voice Extended Markup Language) [8].

The MRF can be split into parts as follows:

- The media resource function processor (MRFP) implements all of the necessary hardware, such as digital signaling processors (DSPs), to be able to stream announcements, mix conferences, and implement IVR dialogs. The MRFP interfaces to the MRFC using a VoIP signaling protocol such as H.248.
- The media resource function controller (MRFC) interfaces to the rest of the IMS network using SIP. It services SIP requests for announcements, conferences, and IVR dialogs. In turn the MRFC instructs the MRFP of the RTP connections and media to stream.

In many implementations the MRFP and the MRFC are collapsed into one networking device to offer the MRF capabilities. This eliminates the need for the extra H.248 interface between the MRFP and the MRFC.

MGW

The media gateway (MGW) is fundamentally responsible for converting media RTP streams on IP to PCM (pulse code modulation) voice channels on a circuit-switched interface. The MGW can support various circuit switching technologies such as DS1, DS3, and possibly SONET optical interfaces. The capacity of the MG will dictate what circuit-switched physical interfaces are supported. On the IMS IP side an MG usually supports various rates of Ethernet.

The MG can also support transcoding in cases where the encoded voice on IP uses one encoding and the circuit-switched channel uses another. Transcoding will convert the encoding from one format to the other, allowing both ends to communicate using different encoding schemes. For example, an AMR [9] may be used on the IMS RTP side and G.711 [10] is used on the circuit-switched side. MGW can support various protection schemes such as 1+1 or 1XN in order to increase the fault tolerance when equipment or transmission facilities fail.

IBCF

The interconnection border control function (IBCF) is a networking device that is used for traffic that is sent between carriers. For example, when a wireless device finds service on a visited network, its registrations and calls must make their way to the home IMS network. Since the network between the carriers can be over an untrusted IP network, an IBCF is used for firewall protection. Various levels of firewall protection exist to guard against unauthorized access and denial of service attacks. Also, since contention in IP addressing used among different carriers can exist, the IBCF can also support Network Address Translation (NAT) functions. The IBCF will also be responsible for refreshing NAT pinholes. This allows the refresh traffic between carriers to be absorbed by the IBCF without putting an additional burden on the rest of the IMS network.

9.2.3 Home vs. Visited Network Components

In an IMS environment there are times when a wireless device is located outside the coverage of its home network service provider. The design of the IMS network is intended to offer the customer the same service independent of where the device is obtaining access. The point of access can be a fixed broadband connection, a femtocell access connection, or a mobile wireless connection. The IMS terminal always registers and processes calls ultimately with its home IMS network. Figure 9-2 shows the major IMS components that are used when an IMS terminal accesses the network from a visiting network.

The section "IMS Flows" provides details on registration and a basic call scenario of an IMS terminal connected to a visiting network. The following lists the major IMS components of the visiting network that provide a mechanism for an IMS terminal to gain access to its home network.

- P-CSCF (Proxy CSCF) is used in the visitor's network as the first device the terminal communicates with when the IMS terminal sends a registration message. The IMS terminal sends the registration message to the visited P-CSCF using the IP provided by the visited network attachment procedures. The SIP request URI in the Register message contains the domain name of the home network where the registration should be forwarded.
- DNS (Domain Name Server) is used to obtain the IP address to which to forward the registration messages based on the domain name from the SIP URI. Domain Name Servers are used by all servers in the IMS network to translate domain and host names to IP addresses to be used to forward messages.

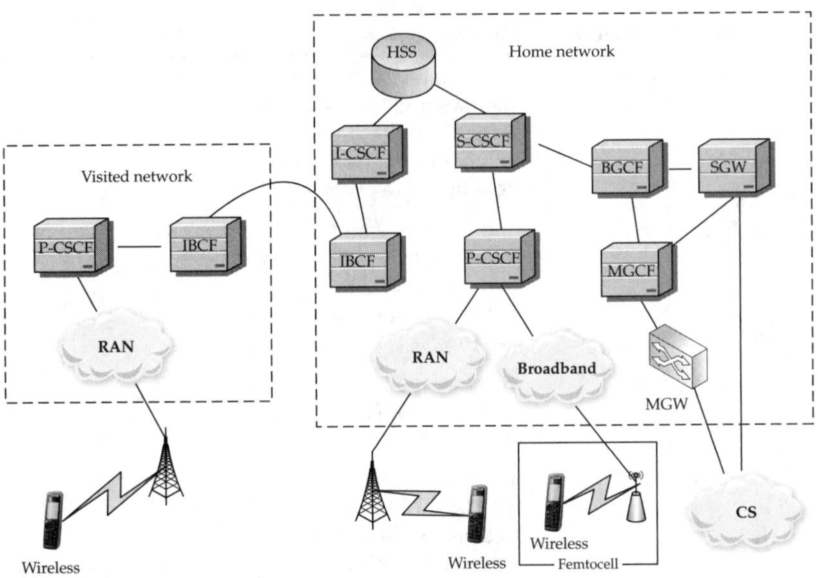

FIGURE 9-2 Home vs. visited IMS components

9.3 IMS Interfaces

The IMS architecture has many components, with each providing a different functionality. The communications among each of the components are identified as well-known interfaces. Sometimes these interfaces are referred to as reference points. The following table lists the more widely used reference interfaces within an IMS network.

Interface	IMS Components	Description	Protocol
Mw	P-CSCF, I-CSCF, S-CSCF	This handles the communications between the various CSCFs.	SIP
Isc	S-CSCF, I-CSCF, application servers	The application servers and the various CSCFs use this interface to exchange signaling messages.	SIP
Mg	MGCF, I-CSCF	The I-CSCF uses this interface with the MGCF for sessions that traverse the PSTN. The MGCF will map SIP sessions to SS7 ISUP sessions.	SIP

Interface	IMS Components	Description	Protocol
Gm	IMS endpoint, P-CSCF	This is the interface between the IMS endpoint and the P-CSCF. The P-CSCF is the only component the endpoint communicates with.	SIP
Mi	S-CSCF, BGCF	The S-CSCF uses this interface to send messages to another IMS network outside of the home network.	SIP
Mk	BGCF, BGCF	This is used to communicate between BGCFs residing in different IMS networks.	SIP
Mj	BGCF, MGCF	The BGCF uses this interface with the MGCF to signal sessions that traverse the PSTN.	SIP
Mm	I-CSCF, S-CSCF, external IP network	This handles communications between an IMS network and an external IP network.	SIP
Mr	S-CSCF, MRFC	This is used to communicate between S-CSCF and MRFC.	SIP
Mn	MGCF, MGW	The MGCF controls RTP and TDM channel mapping on the MGW.	H.248
Mp	MRFC, MRFP	This is used to communicate between MRFC and MRFP.	H.248
Sr	MRFC, application servers	This is used by MRFC to fetch scripts and web pages such as VXML scripts for execution of IVR dialogs.	HTTP and VXML
Cx	I-CSCF, S-CSCF, HSS	This is used to communicate between I-CSCF/S-CSCF and HSS to obtain routing and provisioning data.	Diameter
Dh	Application servers, OSA, SCF, IM-SSF, HSS	Application servers use this interface to locate a HSS in a network that has multiple HSSs.	Diameter
Dx	I-CSCF, S-CSCF, SLF	I-CSCF and S-CSCF use this interface to locate an HSS in a network that has multiple HSSs.	Diameter

Interface	IMS Components	Description	Protocol
Gq	P-CSCF, PDF	This is used to communicate policy-related data between P-CSCF and PDF.	Diameter
Rf	All CSCFs, BGCF, MRFC, MGCF, AS	This is used to provide billing records based on starting, stopping, and feature usage.	Diameter
Ro	AS, MRFC	This is used to exchange online charging information with ECF.	Diameter
Sh	SIP AS, OSA SCS, HSS	This is used to obtain database information between application servers and the HSS.	Diameter

9.4 IMS Flows

In a SIP and IMS network as endpoint devices attach and place calls, protocol messages are exchanged throughout the network. The type of call being placed and the configuration of the network will impact how SIP messages flow through the network. This section presents registration and call flow scenarios along with associated descriptions of the message flows.

9.4.1 Registration Procedure

Figure 9-3 shows an IMS terminal registration. In this scenario the IMS terminal is connected to a visited network.

The following provides a description of the messages exchanged in Figure 9-3 for an IMS terminal registration.

1. When an IMS terminal enters a visiting network, it needs to register itself with its home S-CSCF to obtain service. The IMS terminal sends a registration message to the visited P-CSCF using the IP address provided by the visited network–attachment procedures. The SIP request URI in the Register message contains the domain name of the home network of where the registration should be forwarded.
2. The P-CSCF in the visited network queries the DNS server to translate the domain name of the home I-CSCF to an IP address. The DNS in the visited network will store one or more I-CSCF entries for each domain name on whose basis it needs to be able to forward SIP messages to other home networks.

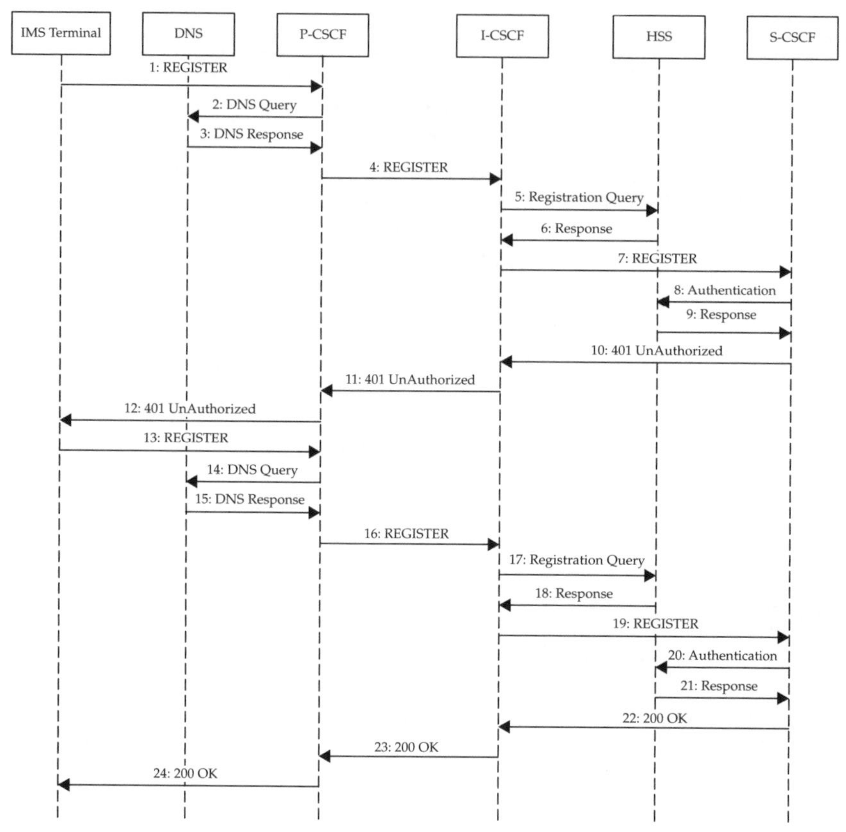

Figure 9-3 IMS terminal registration

3. If an entry exists in the DNS server for the home network's domain name, one or more IP addresses are returned. One address for each I-CSCF in the home networks is returned in the response from the DNS server.
4. The P-CSCF forwards the register message to the I-CSCF of the home network based on the IP address and port provided by the DNS server.
5. The home I-CSCF queries its HSS to find out which S-CSCF is assigned to the registering IMS terminal. The HSS can have a dedicated S-CSCF for each IMS terminal, or it can implement an assignment algorithm that allows for improved load balancing and higher service availability.
6. The HSS responds with the assigned S-CSCF for the registering IMS terminal.
7. The I-CSCF forwards the Register message to the S-CSCF assigned by the HSS.

8. The S_CSCF queries the HSS for the registering IMS security credentials and other important provisioning information.
9. The HSS replies to the S-CSCF with the authorization needed to be provided by the registering IMS terminal.
10. Since this is the initial registration by the IMS terminal, the S-CSCF needs to authenticate the registering terminal. The S-CSCF sends a challenge 401 message, containing an authentication vector, back toward the registering IMS terminal.
11. The home I-CSCSF receives the 401 from the S-CSCF and forwards it back to the visiting P-CSCF.
12. The visiting P-CSCF receives the 401 challenge from the S-CSCF and sends it to the registering IMS terminal.
13. The registering terminal receives the 401 challenge and the IMS terminal resends the register, this time containing the response to the authentication challenge.
14. Again when the P-CSCF receives the registration message, it queries the DNS server for the home I-CSCF address.
15. When the DNS server returns the IP address of the home I-CSCF, the P-CSCF forwards the registration to it.
16. The P-CSCF forwards the register message to the I-CSCF of the home network based on the IP address and port provided by the DNS server.
17. Again the home I-CSCF queries its HSS to find out which S-CSCF is assigned to the registering IMS terminal.
18. Again the HSS responds with the assigned S-CSCF for the registering IMS terminal.
19. Again the I-CSCF forwards the Register message to the S-CSCF assigned by the HSS.
20. Again the S-CSCF queries the HSS for the registering IMS security credentials and other important provisioning information.
21. The HSS replies to the S-CSCF with the authorization needed to be provided by the registering IMS terminal.
22. This time the authorization information contained in the registration is now compared to the authorization information returned by the HSS. If the authorization matches, the registration is accepted by the S-CSCF. The S-CSCF accepts the registration this time by sending a 200 OK back toward the I-CSCF.
23. The I-CSCF receives the 200 and forwards it back to the visited P-CSCF.
24. The P-CSCF sends the 200 OK back to the registering IMS terminal. At this point, the IMS terminal is successfully registered.

9.4.2 Preconditions

As a SIP call is established in a VoIP network, voice path resources need to be in place before an end-to-end conversation can take place. Each device and network element in the voice path can have resources that need to be allocated. In the case of an endpoint device, the codec and RTP ports need to be configured. Network elements such as media gateway devices or P-CSCFs need to have their hardware configured to cut through RTP media. If these resources have not been allocated at the time when a user is alerted, voice path problems can occur. The most likely problem is that a noticeable delay occurs from the time the called party answers and the time the voice path is cut through. Users expect to have an end-to-end voice path as soon as they answer a phone that is ringing.

To address this problem, RFC 3312 [11] was created, which is called "Integration of Resource Management and SIP." This RFC defines the concept of preconditions, which expands on the SDP offer/answer model. As a SIP session is being established, the SDP offer/answer is used first to reserve resources before the called party phone rings. To express the preconditions using an SDP encoding, additional attributes have been defined for the desired and current QoS state. The local or remote QoS state takes on values such as send/receive, send only, receive only, and none. These state values reflect the device's resource ability, current or desired, to pass voice RTP traffic in the direction specified. The RFC offers a generic resource reservation framework using preconditions as needed.

The called user equipment when first sending a response to an INVITE indicates if a QoS commitment is required for the call. If so, then the offer/answer SDP exchange is required to reserve resources prior to the called party being allowed to ring. The IMS Basic Call and IMS to Circuit Switch call flows provide examples of these precondition usages.

9.4.3 IMS Basic Call

Figure 9-4 shows an IMS terminal initiating a call while connected to a visitor's IMS network.

Invite Flow from UE1 to UE2

When a user decides to call someone, she will enter the phone number of the person she wants to reach using the keypad on her IMS phones. On some phones the user will first hear a local dial tone generated by the phone, indicating to the user that it is ready for the user to enter a phone number to call. The following is a description of the messages exchanged to initiate the call setup from UE1 to UE2.

1. User Endpoint 1 (UE1), after collecting all of the digits from the user to dial, sends a SIP INVITE to the visiting P-CSCF. UE1 sends the INVITE and all messages it exchanges with

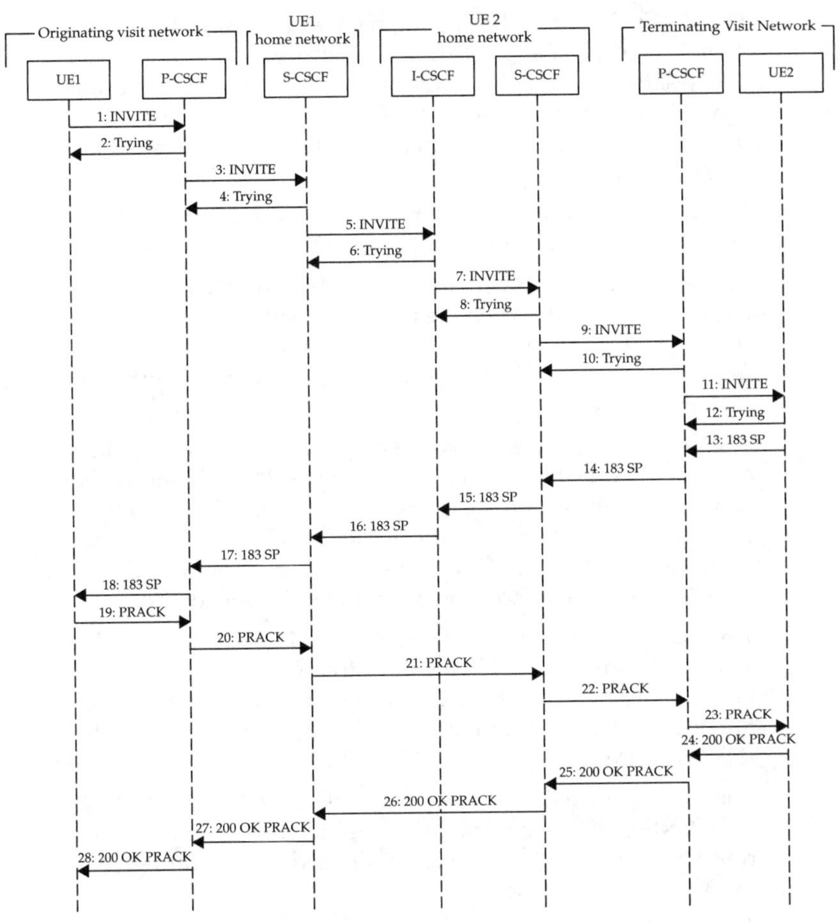

FIGURE 9-4 IMS terminal initiating a basic call

the visiting network to the P-CSCF that was discovered during registration. The request URI contains the telephone number of UE2. Included in the INVITE is the SDP offer that specifies the type of codec(s) and bandwidth being requested for the call. This offer can be a multimedia connection request whereby more then one media stream is being requested such as voice and video. For each media stream being requested, one or more proposed codecs will be listed. Embedded in the SDP are the IP address and ports where the far end (UE2) will stream packets to once the connection has been fully negotiated. Also included in the SDP are the QoS precondition variables that indicate the current and desired resource settings.

2. The P-CSCF sends a 100 Trying back to UE1 to indicate that it has received the INVITE and is processing the request. When UE1 receives this message, it stops a retransmission timer. If for some reason the INVITE or the 100 trying is lost, the timer running on UE1 will expire and UE1 will resend the INVITE.
3. The P-CSCF inserts itself in the record route header and the via header. It then forwards the INVITE to the home S-CSCF.
4. The S-CSCF acknowledges the INVITE by sending a 100 Trying back to the P-CSCF. When the P-CSCF receives this message, it stops a retransmission timer. Again if the INVITE or the 100 trying is lost, the timer running on the P-CSCF would expire and it would resend the INVITE.
5. The S-CSCF performs an ENUM number translation and then queries its home HSS to find out where the INVITE should be forwarded to based on the translated number. In this example the number translation results in the INVITE being forwarded to the I-CSCF of the UE2 home network. The S-CSCF inserts the called telephone number on the PAI header.
6. The I-CSCF of the UE2 home network acknowledges the INVITE by sending a 100 Trying back to the S-CSCF of the UE1 home network. Again the INVITE and the 100 Trying transaction are protected by a timer started on the S-CSCF when the INVITE was sent.
7. The I-CSCF for the UE2 home network forwards the INVITE to UE2's S-CSCF. The I-CSCF doesn't insert itself in the record route header, since further signaling need not go through the I-CSCF for this session.
8. The S-CSCF of UE2's home network acknowledges the INVITE by sending a 100 Trying back to the I-CSCF of the UE2 home network.
9. The S-CSCF of UE2's home network forwards the INVITE to the P-CSCF of the terminating visitor's network. The S-CSCF uses the UE2's contact information, obtained during its registration, to forward the INVITE.
10. The P-CSCF of the terminating visitors network acknowledges the INVITE by sending a 100 Trying back to S-CSCF of UE2's home network.
11. The P-CSCF of the terminating visitor's network forwards the INVITE to UE2. The INVITE will contain information for UE2 to present to the called user such as caller ID. Also included in the INVITE is the offer SDP from the originator.
12. The UE2 of the terminating visitor's network acknowledges the INVITE by sending a 100 Trying back to P-CSCF of the terminating visitor's network.

183 Session Progress

Session Progress 183 is used to provide an SDP answer in a response message to the INVITE from UE2 to UE1. This section describes the 183 Session Progress message exchange.

13. After UE2 processes and decides to accept the INVITE, it builds a 183 Session Progress Response message. The 183 message is sent back to the P-CSCF of the terminating visitor's network. The 183 contains an answer SDP including the chosen set of codecs supported by UE2. The INVITE may have contained one or more codecs for each media stream, of which UE2 picks one codec for each stream. Typically, a phone will have provisioned an order of precedence of which codecs are preferred. Also included in the SDP are the QoS parameters acceptable by UE2. The final QoS agreement can occur either in this 183 or in the answer 200 in Step 24. In this call flow UE2 requests a QoS confirmation using preconditions that will require an UPDATE message to confirm resources are allocated end to end before UE2's phone rings.
14. The P-CSCF of the terminating visitor's network forwards the 183 back to UE2's home network's S-CSCF. The P-CSCF can decide to insert the UE2 URI in the P-Assert-Identify header.
15. UE2's home network's S-CSCF forwards the 183 back to its home network's I-CSCF. The I-CSCF inserts UE2's Public URI in the P-Asserted-Identifier to be provided back toward UE1.
16. The home network's I-CSCF forwards the 183 to the UE1's S-CSCF.
17. UE1's S-CSCF forwards the 183 back to the originating network's P-CSCF.
18. The originating network's P-CSCF forwards the 183 to UE1. UE1 now has the answer SDP sent from UE2.

PRACK/200 OK Exchanged

When UE1 receives the 183, it typically will contain a 100 rel Require header. The 100 rel tag indicates that an acknowledgment of this response has been requested. UE1 receives UE2's chosen codecs in the 183 response.

19. UE1 will send a PRACK message to acknowledge the 183 response message back to the originating network's P-CSCF. UE1 has an opportunity to update its SDP in the PRACK it sends back toward UE2. UE1 at this point will provide a new SDP offer that will have narrowed down the codec chosen for each stream based on the codec response received in the 183.

20–23. These messages forward the PRACK back towards UE2. UE1 will also start to reserve resources now that the scope of codecs has been narrowed down. UE1 will include in the SDP the current QoS reservation status.

24. After UE2 receives the PRACK, it acknowledges it by sending a 200 OK back toward UE1. At this point UE2 will also start to allocate resources now that the scope of the codecs has been narrowed down. The 200 OK to the PRACK contains the answer SDP and the QoS status is updated.

25–28. The 200 OK for the PRACK is forwarded back through the network to UE2.

Figure 9-5 shows the resource reservation and call completion for a basic call. The following provides a description of each of the messages in Figure 9-5.

29. At some point the resources have been completely allocated, including any radio bandwidth allocations that are necessary. In message 13 UE2 requested that the QoS resources be fully confirmed. As a result UE1 sends an UPDATE message to explicitly indicate that all resources have been allocated.

30–33. The UPDATE message makes its way through the network to UE2.

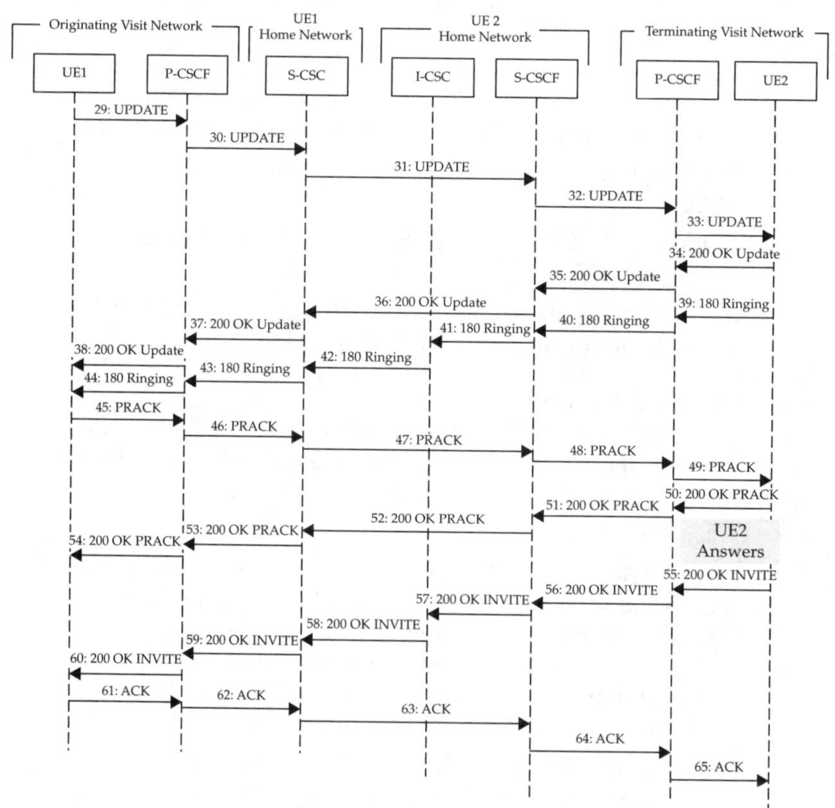

Figure 9-5 IMS basic call reservation and completion

34. When UE2 receives the UPDATE indicating that all resources have been allocated on UE1's end, it acknowledges the UPDATE by sending a 200 OK. Included in the 200 OK is UE2's QoS status, which at this time will most likely indicate that all resources have been allocated.

35–38. The UPDATE messages progress through the network from UE2 to UE1.

9.4.4 Ringing/PRACK/200 OK Exchanged

39. After UE2 has completed reserving resources for all media streams, it can now ring the phone to alert user 2. Note that UE1 has indicated that it has already finished reserving resources in the UPDATE message. In order to inform UE1 that user 2's phone is ringing, it sends a 180 Ringing message to the terminating network's P-CSCF back toward UE1.

40–44. The 180 Ringing messages progress through the network from UE2 to UE1.

45. When UE1 receives the 180 Ringing, it applies a ringback tone to the speaker/earpiece to let user 1 know that the person she called is being alerted. UE1 informs UE2 that it received the 180 Ringing by sending a PRACK message.

46–49. The PRACK messages progress through the network from UE1 to UE2.

50. When UE2 receives the PRACK request, it acknowledges it by sending 200 OK back toward UE1. The 200 OK message contains a Cseq number that matches the PRACK received and the method in the Cseq is indicated as PRACK.

51–54. The PRACK messages progress through the network from UE2 to UE1.

55. User 2 decides to answer his phone after hearing it ring. This causes UE2 to send a 200 OK response with the Cseq number that matches the original INVITE.

56–60. The 200 OK INVITE response messages progress through the network from UE2 to UE1.

61. Once UE1 receives the 200 OK for the INVITE, the call is consider active and connects its codec to the earpiece and microphone. UE1 begins streaming RTP voice packets to UE2. In order to indicate the 200 OK was received, UE1 sends an ACK message back toward UE2.

62–65. The ACK messages progress through the network from UE1 to UE2. Once UE2 receives the ACK from UE1, it considers the call to be active and connects its codec to the earpiece and microphone. UE2 begins streaming RTP voice packets to UE21. A two-way audio connection now exists between UE1 and UE2.

IMS to Circuit Switch Basic Call

An IMS user decides to call someone who happens to be serviced by a conventional circuit-switched network using a local POTS service provider. Figure 9-6 shows the call flow of an IMS terminal placing a call to a circuit-switched customer. In this scenario the IMS endpoint (UE1) initiates the call from within its home network, which contains a P-CSCF, an S-CSCF, a BGCF, an MGCF, and an MGW. The circuit-switched network is shown with a circuit switch that is directly connected to the POTS telephone (UE2).

The following sections provide a description of the messages exchanged in the call flow provided in Figure 9-6.

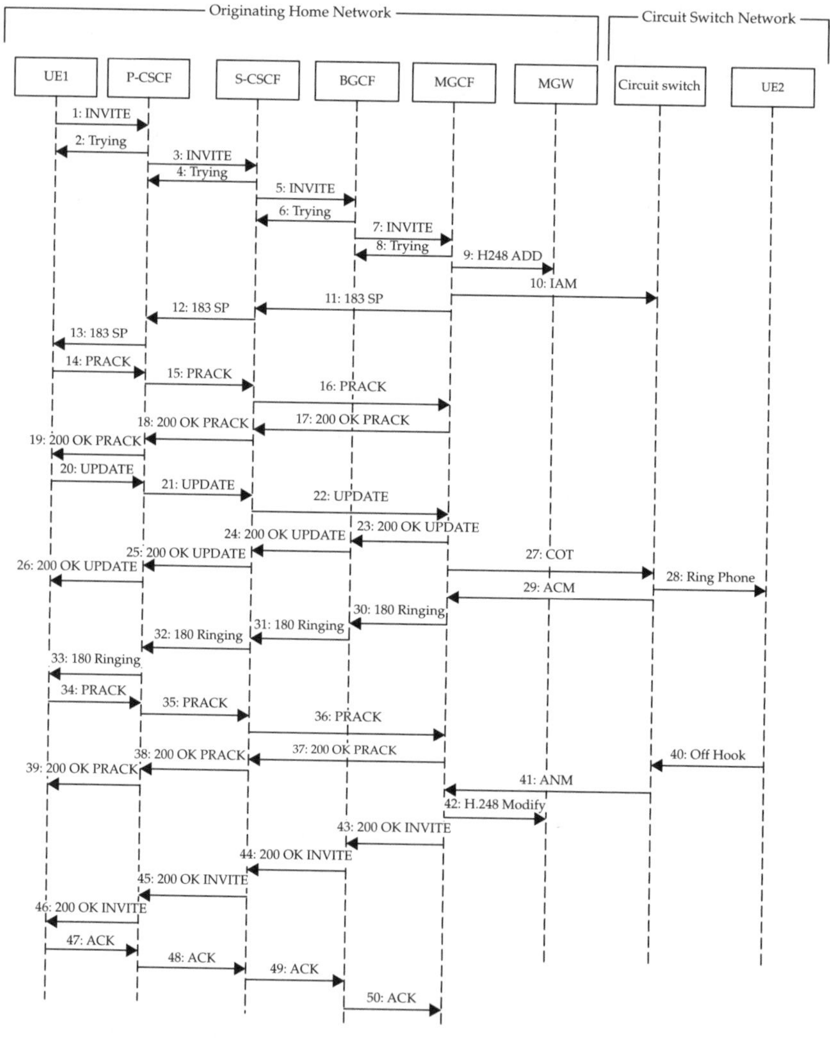

FIGURE 9-6 IMS to circuit switch call

Invite Flow from UE1 (IMS) to UE2 (POTS)

The following describes the initial INVITE:

1. User Endpoint 1(UE1), after collecting all of the digits (E.164 telephone number) from the user to dial, sends a SIP INVITE to the UE1's home P-CSCF. UE1 sends the INVITE and all messages it exchanges to its home's P-CSCF that was discovered during registration. The request URI contains the telephone number for UE2. Included in the INVITE is the SDP offer that specifies the type of codec(s) and bandwidth being requested for the call. The offer can be a multimedia connection request whereby more than one media stream is being requested such as voice and video. In this case, since the person being called is on a POTS network, the final SDP negotiated will be a single voice codec. For each media stream being requested, one or more proposed codecs will be listed. Embedded in the SDP are the IP address and ports where the MGW will stream packets to once the connection has been fully negotiated. Also included in the SDP are the QoS precondition variables that indicate the current and desired resource settings.
2. The P-CSCF sends a 100 Trying back to UE1 to indicate that it has received the INVITE and is processing the request. When UE1 receives this message, it stops a retransmission timer. If for some reason the INVITE or the 100 Trying is lost, the timer running on UE1 will expire and it will resend the INVITE.
3. The P-CSCF inserts itself in the record route header and the via header. It then forwards the INVITE to its home S-CSCF.
4. The S-CSCF acknowledges the INVITE by sending a 100 Trying back to the P-CSCF. When the P-CSCF receives this message, it stops a retransmission timer. Again, if the INVITE or the 100 Trying is lost, the timer running on the P-CSCF will expire and it will resend the INVITE.
5. The S-CSCF performs an ENUM number translation and then queries its home HSS to find out where the INVITE should be forwarded to based on the translated number. In this example the number translation results in the INVITE being forwarded to the BGCF (breakout gateway control function) of the home network to establish a PSTN call. The S-CSCF inserts the called telephone number on the PAI header and inserts itself in the P-Charging Header for billing purposes.
6. The BGCF of UE1's home network acknowledges the INVITE by sending a 100 Trying back to the S-CSCF of UE1's home network. Again the INVITE and the 100 Trying transaction are protected by a timer started on the S-CSCF when the INVITE was sent.

7. The BGCF forwards the INVITE to the MGCF (media gateway control function). The MGCF is responsible for controlling the media gateway for terminating the RTP stream and switching it to a circuit-switched channel such as a DSO in a T1 carrier.
8. The BGCF acknowledges the INVITE by sending a 100 Trying back to the I-CSCF of UE2 home network.
9. The MGCF sends a H.248 Add command to the MGW to reserve a RTP termination in preparation for the RTP to circuit-switch conversion.
10. The MGCF sends a SS7 IAM message to the PSTN circuit switch to request a call to be established to UE2.

183 Session Progress

Session Progress 183 is used to provide an SDP answer in a response message to the INVITE from the MGCF to UE1. This section describes the 183 Session Progress message exchange:

11. The MGCF sends a 183 session progress back to the home S-CSCF. Session Progress 183 is used to provide an SDP answer in a response message to the INVITE. The 183 contains an answer SDP including the chosen set of codecs supported by the MGW. The INVITE may have contained one or more codecs for each media stream, of which the MGCF picks one codec for each stream. In this case, the MGW may only support voice circuits, and as a result, the MGCF will choose its preferred voice codec. Also included in the SDP are the QoS parameters acceptable by the MGCF. The final QoS agreement can occur either in this 183 or in the answer 200 OK to the INVITE in message 23. In this call flow the MGCF requests a QoS confirmation using preconditions that will require an UPDATE message to confirm that resources are allocated between UE1 and the MGCF.

12–13. These are the 183 session progress messages forwarded from the S-CSCF to the P-CSCF and finally to UE1.

PRACK/200 OK Exchanged

The following describes the PRACK and 200 OK exchange.

14. UE1 will send a PRACK message to acknowledge the 183 response message back to the originating network's P-CSCF. UE1 has an opportunity to update its SDP in the PRACK. UE1 at this point will provide a new SDP offer that will have narrowed down the voice codec received in the 183.

15–16. These messages will forward the PRACK to the MGCF. UE1 will also start to reserve resources now that the specific codec has been chosen. UE1 will include in the SDP the current QoS reservation status.

17. After the MGCF receives the PRACK, it acknowledges it by sending a 200 OK back toward UE1. At this point, the MGCF will also start to allocate resources now that a single codec has been chosen. The 200 OK to the PRACK contains the answer SDP and the QoS, including an updated status.

18–19. These messages will forward the 200 OK for the PRACK to UE1.

20. At some point the resources have been completely allocated, including any radio bandwidth allocations that are necessary. In message 11 the MGCF requested that the QoS resources be fully confirmed. As a result UE1 sends an UPDATE message to explicitly indicate that all resources have been allocated.

21–22. These are the UPDATE messages as they make its way through the network to the MGCF.

23. When the MGCF receives the UPDATE indicating that all resources have been allocated on UE1's end, it acknowledges the UPDATE by sending a 200 OK. Included in the 200 OK is the MGCF QoS status, which at this time will most likely indicate that all resources have been allocated.

24–26. These are the UPDATE messages as they progress through the network from the MGCF to UE1.

27. The MGCF sends a SS7 COT message to the circuit switch to request that the called party be alerted.

28. When the circuit switch receives the COT message, it then rings the UE2 phone.

29. The circuit switch will send a SS7 ACM message back to the MGCF to indicate that the UE2 phone is ringing.

Ringing/PRACK/200 OK Exchanged

The following describes the ringing, PRACK, and 200 OK exchange.

30. When the MGCF receives the ACM, it sends a 180 Ringing message to the BGCF.

31–33. These messages forward the 180 message from the BGCF to UE1. When UE1 receives the 180 Ringing, it applies a ringback tone to the speaker/earpiece to let user 1 know that the person she called is being alerted.

34. UE1 informs the MGCF that it received the 180 Ringing by sending a PRACK message.

35–36. These are the PRACK messages progressing through the network from UE1 to the MGCF.

37. After the MGCF receives the PRACK, it acknowledges it by sending a 200 OK back toward UE1. At this point, the MGCF will also start to allocate resources now that the scope of codecs

have been narrowed down. The 200 OK to the PRACK contains the answer SDP, and the QoS status is updated.

38–39. These are the 200 responses to the PRACK messages progressing through the network from MGCF to UE1.

40. After UE2 has been ringing for a while, the called user will attempt to answer the call by going off-hook.

41. As a result of receiving the off-hook event, the circuit switch will stop ringing the phone and send an ANM message to the MGCF, which indicates the call has been answered.

42. The MGCF will send a H.248 modify message to cut through the circuit-switched channel and the chosen RTP termination.

43. The MGCF then sends a 200 OK referencing the original INVITE to signal that the call has been fully accepted.

44–46. These are the 200 responses to the INVITE messages progressing through the network from MGCF to UE1.

47. Once UE1 receives the 200 OK for the INVITE, the call is considered active and connects its codec to the earpiece and microphone. UE1 begins streaming RTP voice packets to the MGW, which cross-connects the stream to the allocated circuit-switched channel. In order to indicate the 200 OK was received, UE1 sends an ACK message back toward the MGCF.

48–50. These are the ACK messages progressing through the network from UE1 to the MGCF.

9.4.5 Sample SIP Register Message

The REGISTER message is the first message sent by a device destined to its home S-CSCF. The contents of the REGISTER message primarily provide location information on where the device can be reached in the event that a session needs to be established to it.

The following briefly describes some of the fields shown in Figure 9-7.

- Request URI is the first line of the message, with the method set to REGISTER.
- The To header specifies the AOR (Address of Record) of the device that is registering.
- The From header specifies the AOR of the device responsible for the registration. Typically, this is the same value as the To header except for scenarios that use third-party registrations.
- The Expires header contains the amount of time in seconds the registration should be valid for. In this case 7200 seconds is two hours before the registration expires and a refresh registration will need to be sent.

```
REGISTER sip:registrar.femtocell.example.com SIP /2.0
Via: SIP/2.0/UDP femtocell.example.com:5060
Max-Forwards: 70
From:<sip:user@femtocell.example.com>; tag=654321
To: :<sip:user@femtocell.example.com>
Contact: <sip:user@femtocell.example.com>
Expires: 7200
Call-ID: 741943881@femtocell.example.com
CSeq:102 REGISTER
Allow:INVITE,ACK,CANCEL,BYE,REGISTER,SUBSCRIBE,NOTIFY,PRACK,UPDATE
Path: <sip:proxy@pcscf.visitor.com:5060;lr>
Content-Length: 0
```

FIGURE 9-7 Sample SIP registration message

- The call ID uniquely identifies the session. The call ID is generated by the originator of the REGISTER message.
- The Cseq number is used to identify the transaction. It contains the method of the transaction, in this case REGISTER, and also a number to identify the specific REGISTER in the event more than one is outstanding.
- The Allow header lists the SIP methods support by the source of the registration.
- The PATH header is used to list the address of each SIP component the registration traverses as it makes its way to the home S-CSCF.

9.4.6 Sample SIP INVITE Message

The INVITE is the message a SIP device will generate when it wants to establish a multimedia session. Contained in the INVITE are the details of the desired connections to establish, including the destination and type of session (voice, video . . .).

The following briefly describes some of the fields shown in Figure 9-8.

- Request URI is the first line of the message, with the method set to INVITE.
- The From header identifies the device that is placing the call, also known as the calling device.
- The To header specifies the AOR (Address of Record), which identifies the desired destination for the SIP session.
- The call ID uniquely identifies the session. The call ID is generated by the originator of the INVITE message.
- The Cseq number is used to identify the transaction. It contains the method of the transaction, in this case INVITE, and also a number to identify the specific INVITE in the event more than one is outstanding.

```
INVITE tel:+732-555-1212 SIP/2.0
Via: SIP/2.0/UDP femtocell.example.com:5060
Max-Forwards: 70
From:<sip:user@femtocell.example.com>; tag=123456
To: <tel:+732-555-1212>
Call-ID: 15234366756723423 6@femtocell.example.com
CSeq:22 INVITE
Require: precondition
Supported: 100 rel
Contact: < sip: user@femtocell.example.com; transport=udp>
Content-Type: application/sdp
Content-Length: (…)
v=0
o=4724627895 4356734876 IN IP4 femtocell.example.com
s=-
c=192.1.1.47
t=0 0
m = audio 2347 RTP/AVP 15 14
a=curr:qos local none
a=curr:qos remote none
a=des:qos mandatory local sendrecv
a=des:qos none remote sendrecv
a=rtpmap:15 AMR
a=fmtp:15 mode-set=0,2,5,7; mode-change-period=2
a=rtpmap:14 telephone-event
a=maxptime:20
```

FIGURE 9-8 Sample voice call INVITE

- This INVITE requires that preconditions be enforced in the SDP and is listed in the Require header.
- The SDP in this INVITE is requesting QoS preconditions and an AMR voice codec connection.

References

[1] RFC 3320 "Signaling Compression (SigComp)", January 2003.
[2] 3GPP TS 29.199-1 "Open Service Access (OSA), Parlay X Web Services", December 2005.
[3] http://java.sun.com/products/jain.
[4] 3GPP TS 23.206 "Voice Call Continuity (VCC) between Circuit Switched (CS) and IP Multimedia Subsystem (IMS)", March 2007.
[5] 3GPP TS 29.078 "Customized Applications for Mobile Network Enhanced Logic (CAMEL)", April 2010.
[6] ITU-T Recommendation Q.700, "Introduction to ITU-T Signaling System No. 7 (SS7)".
[7] ITU-T Recommendation Q.701-Q.705, "Signaling System No. 7 (SS7) – Message Transfer Part (MTP)".
[8] W3C Recommendation, "Voice Extensible Markup Language", June 2007.
[9] 3GPP TS 26.071, "Mandatory Speech Codec Speech Processing Functions; AMR Speech Codec; General Description".
[10] ITU-T Recommendation G.711, "Pulse Code Modulation (PCM) for Voice Frequencies", November 1988.
[11] RFC 3312 "Integration of Resource Management and SIP," October 2002.

Index

S

T

U

V

W

X